IVI SOLPE

NEW PHYSICAL WORLD VIEW

DIGEST OF ARTICLES

ALTASPERA
PUBLISHING & LITERARY AGENCY INC.

© 2023 – **Ivi Solpe**

ISBN **978-1-4466-6249-6**

Published in Canada
by Altaspera Publishing & Literary Agency Inc.

The book contains three articles by the author devoted to the fundamental concepts of modern physics and our worldview: time, force interaction, coordinate system. Much attention is paid to the historical aspect of how people's ideas arose in the course of their work. First of all, it is concluded that the concept of time was introduced artificially into the description of mechanical movement, based on the practical experience of people, the need to streamline their daily life and work activities. Gradually, the concept of time became a fundamental physical quantity. It has entered into all theoretical descriptions of physical processes, such as speed, acceleration, and manifests itself in the coordinate system, as a segment of the path, etc. But is time a real physical substance or a cotinuum? If not, what should our understanding of the world and the possibility of describing it look like? The author gives an unequivocal answer – time, as a physical reality, does not exist! Based on the principle of relativity, a new concept of «body dynamics» is introduced as a number. The coordinate system of Descartes is also the heir to the practical experience of people who perceived the surface of the earth as flat. They did not see themselves as an «observer» of the world from the center of the system. For them, there was always some particular movement of a material object in one direction, which they tried to describe. However, the experience of Michelson-Morley made its own adjustments. Time has become a variable, depending on the speed. As a result, the distance has also become a variable. The formulas for the transition from one coordinate system to another have also become much more complicated. What did the Michelson-Morley experiment prove? Was Einstein right when he introduced the postulate of the constancy of the speed of light and created the special theory of relativity? The experiment showed that our ideas about the coordinate system are erroneous. The author offers a completely new view of the coordinate system, as radial-spherical. In this system, the Michelson-Morley experiment does not have a «negative» result and is described as the usual propagation of light in all directions, with the same speed. Having revealed to the world the law of attraction of two bodies, Newton made a logical error in interpreting his final result of calculations.

For centuries, using the formula in practical calculations and getting results that are quite suitable for practice, people do not even ask the question:

«Was Newton right?»

But a small logical analysis is enough to understand that, the force interaction of two bodies is proportional not to the product of their masses, but to their

SUM. This form of the law also perfectly describes the motion of planets in the circle of the Sun, but there are also some quantitative differences. Modern methods of experimental research make it possible to identify these differences and finally answer all the main questions.

A person must finally understand his place in this world and learn how to correctly display it in theoretical descriptions.

CONTENTS

ANNOTATION

«Physical foundations of mechanical movement
and its description».

The description of any movement is organically connected with the concept of time. Based on the analysis of historical facts, as well as the works of Aristotel, Newton, Einstein and others, it is shown that time and clocks were first introduced into the life practice of people, and then into science, as a reference movement. A fundamental conclusion is made about the nature of time. It does not represent any physical reality and cannot be considered as something different from the ordinary mechanical movement of bodies. On this basis, the concept of the speed of a body is considered, it is indicated why exactly such a formula was adopted to describe the speed at a certain historical stage and what are its shortcomings. Analyzing the process of observation itself, its fundamental moments are distinguished, which are proposed as conditions for determining the relative speed of the movement of bodies. A new characteristic is introduced – «dynamics of relative motion», as a number.

6

«PHYSICAL FOUNDATIONS OF MECHANICAL MOVEMENT AND ITS DESCRIPTION»

INTRODUCTION

All our observations of the surrounding world are connected, first of all, with the mechanical movement of bodies. The very existence of a person, his life and environment are also connected with it. Therefore, the desire to know the movement, its essence, to describe with the help of some formulas, was a historically necessary need. On this path, man has always put himself in the center and his natural desire was to reflect the world in relation to himself. He tried to find such forms that would also allow him to conduct purposeful activity, coordinating it with the phenomena of the external environment. And these forms were found on the basis of practical life activity. Initial concepts such as space and time, length and movement were introduced. They made it possible to fully describe the movement of bodies, indicate the location of a person, and define derivative concepts such as speed, acceleration. A person was able to set his goals and fulfill desires and interests. His whole life became orderly and predictable. Existing and satisfying man for centuries, these concepts and forms became laws that were not subject to any doubt or revision, acquiring the status of an objective physical reality. The further development of these concepts and their mathematical expression was associated only with modifications of the external forms of their notation. But nothing has been done to analyze and reveal their physical essence and correspondence to each other in terms of their internal structure. This can be clearly seen in historical terms on the example of three outstanding thinkers of their time:

Aristotle, Newton, Einstein. The works of each of them are, as it were, a certain milestone of a new qualitative stage in physics and have absorbed all the previous experience accumulated by mankind. «Physics» of Aristotle is striking in its desire to deeply comprehend the very initial premises and concepts of the theory, to know the essence of the universe through a logical construction. Newton's «Mathematical Principles of Natural Philosophy» is elegant in mathematical construction and is mainly based on the original concepts of Aristotle, such as space, movement and time, giving them some formalism. However, further observations showed that the world is more complex and does not fit into Newton's mathematics. New experimental facts led to the creation of Einstein's theory of relativity. At the same time, the original concepts were again preserved, although with some amendments: space became curved, time acquired the ability to slow down in a moving system. A new concept appeared in the form of Minkowski – «space-time». What is this description of the universe, what is it connected with? Why are there non-linear transformations between two inertial systems moving in a homogeneous and isotropic space?

After all, the whole difference between them represents only one value – the speed of relative movement. And what follows next after this non-linearity? Considering the question of the movement of bodies in a historical aspect, in the course of how one theory, emerging on the foundation of another, leads to a complication of the description of the laws of the universe and an explanation of the basic initial concepts, one involuntarily comes to the conclusion that somewhere in the very beginning a mistake was made. Isn't this connected with the purposeful activity of a person, his desire to streamline the life of a community and describe the world in relation to himself?

Did he introduce some superfluous premises or concepts? If so, then only time could be such a concept. As for other concepts, such as space and movement, they have always existed even before the description of the world by man. But it is precisely the concept of "time" that appeared already as a result of its life activity, at a certain stage of development. By this time, the experience of observing natural phenomena and their relationship with the location of the heavenly bodies had already been accumulated. There was a need to coordinate the social activities of people with the phenomena of nature. And already after the concept of «time» was firmly established and entered into practice and everyday life, it was used to describe the movement of bodies as a very real physical quantity. Was it the right move? Of course, yes, but only from the point of view of the purposeful activity of people as a community, their dependence on natural phenomena. The concept of time and their practical implementation in the clock mechanism made it possible to simply and conveniently streamline all events and describe the movement of bodies relative to a person. There was a unique opportunity to predict certain events and make preparations for them, such as the time of the arrival of the train or the appointment. And it became so important and convenient that the clock became an inapplicable attribute of human existence, and the concept of «time» was firmly established and acquired a halo not only of objectivity, but also of independent physical reality. But is this true in relation to the very physical phenomena of nature, to the movement of matter, which exist independently of the consciousness and will of man? If not, then how would it be physically correct and objective to describe the movement of bodies, what to put in accordance with this process? In this work, answers are given to the «inconvenient» questions of our time.

ON THE «PHYSICAL» NATURE OF TIME

First of all, let's try to trace how the idea of «time» was born in its historical aspect, with the development of human civilization and the accumulation of certain knowledge. This will greatly help to understand the «physical» essence of time and what it is connected with. We will avoid various philosophical theories about the nature of time, which would lead us far away from the essence of the issue. All of them can be divided into two large groups: a) denying the objectivity of the existence of time; b) recognizing the objectivity and reality of the existence of time and investigating its properties. The fact that these theories live and develop for a long time only speaks of the uncertainty of physical science in the question of time. In this regard, Einstein's statement is of interest:

«I am convinced that philosophers have had a detrimental effect on the development of scientific thought by transferring some fundamental concepts from the realm of experience, where they are under our control, to the inaccessible heights of a priori». (10).

When did the first signs of the emergence of the concept of time begin to appear? Apparently, it was connected with the ancient East. Exactly here, thanks to good climatic conditions, for several thousand years BC. in the river valleys and on the coasts of the seas, the first large settlements and cities began to emerge, and then slave-owning states also formed. (1). There is no doubt that, initially, their social activities were determined by the rhythms of day and night. After all, it was necessary to work and rest, and to perform cults, sacred to the gods. At the same time, it was possible to coordinate the joint actions of a large number of people by various methods. These could be biological rhythms and the needs of the person himself, to sound signals and determining the height of the Sun. With the growth of the population and the strengthening of state discipline, these methods were constantly improved and became more and more accurate. This accuracy was enough within the limits of one settlement or district. First of all, man felt and knew the rhythms of nature associated with the movement of the Earth, and tried, as accurately as possible, to follow them within his community. Moreover, it was necessary to somehow take into account long periods, like the change of seasons, which led to the appearance of the calendar. It is believed that the invention of the calendar dates back to ancient times and already three thousand years BC. The Egyptians had a relatively perfect calendar system. (2). Its creators are the ancient Egyptian priests-astronomers. The night, according to this calendar, was divided into 12 parts – hours, and the day was set aside for 10 hours and two parts for twilight, i.e. The day was divided into 24 hours. Let us make a very important remark that the concept of time and the use of clocks for counting have already been widely used. Their year contained 12 months of thirty days each, and on top of that, five holidays were added, which is 365 days in total. From this moment, the so-called "counting time" begins. Those. accounting for the number of certain periods in the phenomena of the surrounding world. The appearance of the calendar, as it were, fixed the discovery of a kind of law on the constancy of these periods, and therefore made it possible to predict them. This became the most important moment in confirming the fact of the reality of time – nature follows its laws. The development of states and trade, the expansion of geography and the success of astronomy put forward increasingly stringent requirements for the accuracy of timekeeping. However, it was this period in the history of physics that turned out to be the least studied. We have not received any data and evidence about the development of physics of that time, about the views on the fundamental concepts that form its basis. (3). This, apparently, is explained by the fact that public life was subordinated to religious rituals. The observed phenomena were explained by the actions of mythical creatures, and the predictions were based on the position of the stars. All astronomical phenomena of that time were carried out for the purpose of astrological predictions, which served as the basis for government decision-making.

Astronomical tables speak of careful and systematic observations of the stars, of attempts to identify mathematical patterns in their movement. Therefore, it is difficult to agree with the fact that the discovery of the coincidence of the beginning of the Nila flood with the summer solstice and the first appearance in the east, before sunrise, of the star Sothis (Sirius), is a harbinger of the birth of the concept of time. (2.4).

Of course, people needed a «signal» about the flood of the Nile, especially since it began suddenly and brought great disasters. But the spill could also be predicted by the calendar. Most likely, the discovery of Sothis was associated with astrology. It is no coincidence that this star was also called the «Star of the Nile». It was really a «heavenly signal», and not a countdown from spill to spill. In their attempts to identify patterns in the movements of the Sun, ancient astronomers came up with the idea of creating a water clock, about two thousand years BC.

They determined the «step» of the Sun, i.e. wanted to establish how many times the path of the Sun along the ecliptic is greater than its diameter. They put this ratio in accordance with the weight of the water flowing out of the vessel. (1). In the future, the unit of time measurement «sussu» was equal to one sixth of a day and was determined by the water clock, as the weight of water flowing out during a given measurement period. (3).

It is difficult to establish the exact date of the invention of the «gnomon» sundial. So, some sources state that the sundial was already known 2500 BC, while others indicate the use of gnomons only 1400 BC. (2.5). The first documentary mention of the gnomon was found in the manuscripts of the Chinese Cheu Pei, approximately 1100 BC. It describes the determination of the height of the Sun above the horizon by changing the shadow from the pole. (1). And only later they began to determine the time by moving the shadow in a circle and applied a scale. Since all astronomical calculations were made in the sexagesimal system, the hours were also divided into 60 minutes. From the second division of the hour, a second was produced. (4). From this moment, the counting of already small periods of time begins, namely, the time that is used in everyday life. It took about a thousand years of development of society and science, understanding the relationship between terrestrial and astronomical phenomena, in order to go from the first «star signals» about spills and the calendar, until the first sundial «gnomons» with a scale.

They were installed on the squares in front of the entrances to the temples and served to honor the cult of the «God of the Sun». Apparently, at this stage in the development of human civilization, a transition was made from counting periods in natural phenomena and their duration, to the emergence of the very concept of time and its constant measurement. Time entered the practice and life of people through natural phenomena, which were repeated with amazing constancy. However, the very concept of time, as a physical quantity or reality, was not subjected to any analysis or philosophical discussions. It served, first of all, religion. Initiated priests skillfully used predictions for their rituals, praising their main gods RA and Osiris. What did «time» mean in that historical era in Arabic? It comes from the root ZMN زمن

12

zaman «time», akin to the word زم zamma «to bind». However, in turn, the Arabic زمن zaman comes from the plural form. numbers أزمنة 'azmina «change», whence it follows that time is change.

In general, we can say that time was presented as a change in interrelated events or natural phenomena. And, if at the beginning only cyclic changes were observed, using natural «signals», then later, with the invention of the clock, individual interrelated events in people's lives began to be measured. A change in one phenomenon was also associated with the clock, over time, for example, human aging. And what is offered to us today to explain the concept of «time»? Here is what Ozhegov writes in his dictionary:

«Duration, the duration of something, measured in seconds, minutes, hours; a certain moment at which something happens».

The meaning remains the same, a certain «duration» is measured with the help of a clock. But why measure it? To commit some change. Time does not produce this change, but the change takes place in nature. Time is only an invented standard, which is compared with the duration of this change. This allows you to get certain experimental results that can be memorized and compared.

As physics has developed, there has been more and more a shift away from observation, external descriptions, and scholastic disputes towards experiment and mathematical proof. For the first time, the movement itself is classified, the concepts of speed and acceleration are introduced and refined.

At the same time, time is successfully used as a value for calculations; it is organically included in the composition of mathematical formulas when describing the mechanical movements of bodies. Conducting experiments required the use of some kind of instruments that would measure objectively occurring physical processes. But a watch is also a physical device that measures «something». Since time is measured by the clock, it follows logically that time is an objective physical reality. It objectively exists independently of our observations and clocks, just like space. And this subtle transition, which «separated» the clock, as a device for measuring, and time, as a physical reality, occurred after

Newton, thanks to advances in the mathematical description of the entire class of movements and the development of philosophical theories. Starting from this moment, the concept of time is already used independently without reference to anything. It extends to all spheres of activity, from everyday life to science, new generations of people are brought up on it, it enters consciousness along with space and matter. The era of a new objective physical reality has come – «time». And throughout the entire period, from antiquity to the present day, disputes about the reality of time, its nature and properties do not subside.

So what does the review of historical facts show us on the way of the origin and establishment of the concept of time as an objective physical reality. First of all, the following points can be distinguished:

a). The emergence of state forms required to synchronize the activities of people, first of all, during the day, as well as for longer periods, it became necessary to predict the phenomena of nature.

в). Astronomical observations and the development of mathematical science made it possible to put forward the idea of a universal standard and embody it in hours.

c). The all-round distribution of time and its successful use in mathematical equations to describe processes in nature and needs in everyday life led to the conviction of its objective existence, as some kind of physical reality. From here, only one fundamental conclusion can be drawn: the origin of the concept of time was associated only with the purposeful activity of a person.

It was introduced by him first into practice and everyday life, and only then into science. The true value of time was in, that it allowed the world to be ordered in relation to the person himself and to synchronize his activity with the phenomena of nature. After all, what is the essence of the clock? Their idea is based on uniform movement, which is divided into separate intervals. For measurements, the clock must be at some point in space, namely in the hands of the observer, so this uniform movement was made circular. From here came the scale and units of time: seconds, hours, and longer periods were counted using a calendar.

Throughout history, man has sought to improve the accuracy of their measurements in order to have the most uniform movement possible, close to ideal. It was this movement that was taken as the standard, relative to which all other movements of bodies, as well as the everyday life of people, are known and ordered. This means that it can be unambiguously determined that the concept of time, in its physical basis, is a standard uniform movement, which was artificially correlated with the clock as a measuring instrument. But does this represent any other physical reality than the movement itself? No, it doesn't, it's the same thing. We have a standard of uniform motion, which we compared with the concept of time, just as we have other standards: a meter, a kilogram, etc.

In Minkowski's view, space and time were not only placed on the same level of physical reality, but also united in one continuum – «space-time». However, this raises a paradox. Space is measured, in general, with a ruler. In this case, the observer, when measuring a segment of the path, must move from the starting point to the end point. When measuring time by the clock, the observer remains in place and fixes the duration of time. He does not need to move anywhere, the movement of the clock itself occurs relative to its rest. But, if time and space are one common physical reality, a continuum, then how do they co-exist. The watch is tied to the ruler and moves along with it. A ruler measures space, but what then is measured by a clock? Imagine, mentally, an observer in a state of weightlessness, in an empty space in which there are no bodies and light. Objectively, he will not be able to determine whether he is moving or not and in what direction. At the same time, he will not be able to determine the length of the path traveled, since it has no beginning and end of the countdown. But the space really exists for him, he stays in it, although the movement for him, as an observer, «froze».

But will he be able to determine his stay in time, if he never knew about its dimensions. In his world there are no cycles of nature, no movement of the Sun and stars. It makes no difference to him whether it is day or night, flood or drought. He will feel only changes within his body, but he does not need to measure time. No one will have any prerequisites for introducing this concept. Perhaps the first attempts to understand the physical essence of time belong to Aristotle. (1.7). In almost all works on the history of physics, his views are analyzed and citations are given for his research. But nowhere is his main thesis about time found, on the basis of which it is necessary to begin to philosophize about the nature of time:

«...what is time and what is its nature, it is equally not clear what is transmitted to us from others, and from what we had to make out earlier».(7).

At the same time, Aristotle very clearly defines two points. Firstly, an assessment is given of all the accumulated experience, in the entire history of the development of science before it. Secondly, he sums up his research and reflections on the concept of time, which did not lead to the disclosure of its physical nature. One can only guess what a painfully difficult situation such a thinker as Aristotle was in, the meaning of his whole life was to know the essence of phenomena. Not understanding the physical nature of time and observing its wide distribution, inalienability from real life and various physical and mathematical descriptions, he was forced to force himself to somehow explain what kind of phenomenon «time» is. Aristotle makes the only possible philosophical transition:

«Time is not movement, but it does not exist without movement».

The fact that time cannot be a movement, he connects with the diversity of the observed movements, they can be slow and fast, moreover, these movements are limited to some area of space, so that there would be a real possibility for their measurement . On the contrary, time must exist everywhere and with everything, and its changes occur only evenly. Having proved that time is not movement, he had, further, to somehow connect it with this movement itself.

Turning to the analysis of «now», Arstotle points out some of its properties:

– «...if "now" were not different every time ... there would be no time»;

– «now» changes time because it precedes and follows».

«It itself, in one respect, is identical, in another it is not, it is different, because it is always in a different and in a different time, on the other hand, «now» is identical in terms of the subject»;

c) «...a moving object is accompanied by «now», as time accompanies movement»;

d) «now» is obviously not a particle of time»;

e) «...since «now» is a boundary, it is not time and is inherent in it by coincidence; because it serves for counting – there is a number».

g) «now» is the end and the beginning of time, but not the same time, but the end of the past and the beginning of the future».

Fundamental in this analysis is the assertion that time, which is not a movement, only accompanies any movement. And, since time is one of the forms of movement, it turns out that a particular movement (standard) accompanies all the variety of movements. It is understandable, because time must exist everywhere and with everything, but how does it «accompany» at the same time. However, Aristotle does not explain how a single «now» can accompany all the diversity of the movement of bodies, from the point of view of physical reality. But this is already a consequence of the «philosophical impasse» and the inability to understand the essence of time, which he directly points out at the beginning. Since the physical process cannot occur simultaneously everywhere, in all space, then time, through its «now», takes on the role of some abstraction, devoid of physical nature.

«...When there is before and after, then we are talking about time, because time is something other than the number of motion, in relation to the previous and the next. Thus, time is not movement, but is it insofar as movement has a number».

And further:

«...we not only measure movement by time, but also time by movement, due to their mutual definition, for time determines movement, being its number, and movement determines time». (7).

And only after time has acquired a mystical «flesh» in the form of a number, which is able to accompany any movement, Aristotel will say:

«It is obvious that every change and every movement takes place in time».

Time is a number in motion – that is the physical nature of time. This was, perhaps, the only possible option for him, how, in general, it was possible to link an abstract concept with a really existing world, while endowing it with certain properties. Obviously, these very skillful philosophical constructions suffer from inconsistency. How can some physical reality accompany any physical process or movement and at the same time contain it in itself?

Other Aristotle's statements about time are also noteworthy:

«...further, as the same movement can be repeated over and over again, so can time, for example, a year, spring or autumn».

Considering the question of what movement is the measure of time, he concludes:

«...neither a qualitative change, nor growth, nor emergence are uniform, but only displacements. That is why time seems to be the movement of a sphere, because other movements are measured by this movement, and time is measured by it»

Is this not direct evidence that time entered practice and science from the awareness of periodicity in natural phenomena and which, in the future, was associated with the daily movement of the Earth. And his words that other movements are measured by this movement speak of his guess, in the choice of a reference movement, as the basis for comparing them with each other.

The enormous influence of Aristotle's ideas on the development of mechanics continued until the 17th century. (3). Various views on the idea of time were expressed by other thinkers, but they did not have any influence on the description of movement, and also could not reveal the very nature of time. (1,3,6). In essence, the views of Aristotle were replaced by Newton's mechanics, his ideas and views on space and time. By this period, ideas had already begun to take hold that space and time exist independently of matter, on their own.

From these positions, Newton simply postulates the concepts of absolute space and time. (8).

However, he does not, in any case, on the pages of his famous «Beginnings», attempts to understand the essence of these concepts. Why did it happen?

Newton saw his main task in subordinating the phenomena of nature to the laws of mathematics. Following this principle, he writes in the introduction:

«We propose this essay as the mathematical foundations of physics. The whole difficulty of physics, as will be seen, consists in recognizing the forces of nature from the phenomena of motion, and then explaining the rest of the phenomena by these forces».

The problem was, first of all, to find an image of a reference system in which the law of inertia would be fulfilled. And, as a physicist, he well understood that reference systems associated with material bodies could never become the basis for a mathematical law. When they move in space, some perturbations are always introduced. If he would choose the movement of the Earth as time, then the law of inertia would not be valid, despite the fact that there are some irregularities in this movement. Arguing in this way, he came to the image of absolute time. But it was just a mathematical image, and not the very concept of time, as a physical reality. Assuming, thereby, an absolutely uniform course of time, Newton got the opportunity to recognize through these mathematical laws arbitrarily small changes in the motion of real bodies, as well as the forces acting on them.

About time, Newton wrote:

«Absolute, true or mathematical time, in itself and by virtue of its internal nature, flows in the same way, without regard to anything external, and is otherwise called duration».

It is unlikely that Newton would have derived his laws of mechanics if he had delved into the analysis of the essence of the concept of time, but simply had not introduced a kind of «standard of time». He directly points to this:

«Perhaps there is no such thing as a standard movement by which time can be accurately measured.»

In some way, Newton's absolute time resembles Aristotle's time, as «a number in motion». Neither one nor the other is reproducible in nature and exists only as an abstract image that helps to describe physical phenomena.

But, if Aristotle offers a purely qualitative, philosophical picture of the world and the problem of the accuracy of calculations does not affect him, then Newton already distinguishes between mathematical and ordinary time:

«Relative, apparent or ordinary time is either exact or changeable, comprehended by the senses, external, performed through any movement, a measure of duration used in everyday life, instead of a true mathematical one, such as an hour, a day, month year». (8).

Again, there is no indication of the nature of time. It is difficult to combine such different concepts as «sensual» and «external». After all, we associate the concept of «feeling» with a person, therefore, from this point of view, time is subjective. But, on the other hand, «external» means just the existence of some physical phenomenon independent of the consciousness of the subject. But how is it possible to determine the measure of «duration» through movement without a subject? These questions, as well as the duality in the definition of the concept of time, caused numerous disputes and gave impetus to philosophical theories. However, physicists accepted Newton's position, distinguishing between absolute and relative time. (1). They were quite satisfied with the correspondence of the derived laws, on the basis of the established axioms, to the experimental data. Time firmly entered the equations of physics as an independent variable, and this further strengthened faith in its objectivity and reality. But without revealing the physical essence of time, Newton practically identified it with movement, since he determined duration through «any movement». And, if for Aristotle, time «seems to be the rotation of a sphere», then for Newton, it is already «performed through movement». His main merit in this direction was to find a mathematical model of the reference movement, which could be put in line with the concept of time. Not finding this standard in nature, Newton introduces the image of absolutely uniform motion, which he defined as mathematical time. Time has lost its physical nature and has become a purely abstract quantity. But physical formulas appeared, the results of calculations according to which, approximately, described the movement of bodies. Thus, a certain matrix was created, which was superimposed on natural phenomena.

Further development of the concept of time is associated with new experimental data and the creation of the foundations of the special theory of relativity (SRT). (1,3,6,9). The Michelson-Morley experiment showed the independence of the speed of light from the movement of the source, did away with the idea of the ether and established the impossibility of determining the absolute movement.

Having connected the Lorentz transformations with the contraction of a moving body and its local time, Einstein transferred them to the properties of space-time and created the foundations of a new kinematics. In doing so, he introduces two postulates:

– about the principle of relativity;
– on the constancy of the speed of light.

In his first work, On the Electrodynamics of Moving Bodies (1905), he approaches the concept of time through an argument about simultaneous events. (9). Using a thought experiment with clocks, he arrives at the following definition of time:

«The time of an event is, simultaneously with the event, the indications of a clock at rest, which are located at the place of the event and which run synchronously with some clock at rest. According to experience, we also believe that the value:

$$2AB \: / \: (t' - t) = V = c$$

is a universal constant (the speed of light in a vacuum)».

Entirely and completely, linking the concept of time with the hands of the clock, he departs from clarifying the nature of time. Instead, there is the problem of ensuring that the clock runs in sync. In what follows, Einstein's approach is purely formal. He focused on the mathematical derivation of the invariant equations of electrodynamics using Lorentz transformations. Only in his third work «On the principle of relativity and the consequences arising from it» Einstein deduces the basic relations of the dynamics of the motion of a body. But here, as in other works, we do not find an analysis of the concept of time and its properties as some kind of physical reality. Instead, the issue of simultaneity and synchronization is analyzed in great detail, as before, using the hands of a clock. (10). This happened for the following reasons. It is believed that absolute time does not exist and the assertion that the event A occurred simultaneously with the event B does not make sense, since there are no means to prove it. After all, real clocks with arrows, with which time is associated in SRT, have a running error, which certainly increases. The question arises, is there a way by which it is possible to eliminate this error, i.e. synchronize clock. Moreover, the clocks are in different inertial frames and, in the general case, do not rest relative to each other.

It is quite natural that it was proposed to use a light signal for this purpose. After all, it was he who was credited with the unique ability to keep his speed unchanged in any inertial frame, especially since this was proved by experiment. Based on this, another well-known conclusion is made that the concept of simultaneity is a relative concept, and, therefore, each inertial frame has its own time.

Further, developing the ideas of SRT, Minkowski will say in his report:

«The object of our perception is always only places and times taken together. No one ever notices a place except at a certain time, or at a time other than in a certain place. A point in space at a point in time, i.e. the system of values X;Y:Z;t I will call the world point». (9).

He points out that all his views are based on experimental data and mentions, first of all, the Michelson-Morley experiment. Taking, at the very beginning of the research, the abstract positive parameter "**c**", he considers a geometric figure, which is described by the equation:

$$c^2t^2 - x^2 - y^2 - z^2 = 1$$

Analyzing this equation, he gives a physical meaning to the parameter "**c**" as the value of the speed of light in vacuum and gets his famous «space-time» invariant. Noting its versatility in deriving the equations of physics, Minkowski concludes:

«From now on, space by itself and time by itself must turn into fictions, and only some kind of combination of both must still retain independence». (9).

It is impossible, of course, to give an analysis of the concept of time, if it is a fiction, or to prove its physical reality. Even to his own construction of «space-time», he assigned the role of «some connection», without revealing its physical nature.

There is some analogy in the concepts of time, which were used in the description of physical processes by Newton and Einstein. Newton distinguished between absolute and relative time.

He believed that in nature there is no process that could reproduce absolutely uniform motion.

In his laws, he used only mathematical time, i.e. abstract value. Einstein, relying primarily on the result of the Michelson-

Morley experiment, and postulating the constancy of the speed of light in a void, had at his disposal not only absolute uniform motion, but also its constant speed relative to any inertial systems. In essence, he had a universal reference movement in any system, against which it is possible to compare and adjust other real movement, which we associate with hours and time. Having such a standard, Einstein introduces it into all his equations of kinematics. And again, this is purely formal, like a mathematical device.

But any standard serves, first of all, to detect an error. This is exactly how Einstein considered the idea of using a light signal to synchronize clocks in different systems. However, in the SRT equations, the principle of clock synchronization in its pure form is absent. Two parameters appeared: time, as a variable, and the speed of light, as a standard. The standard connects all inertial systems and synchronizes time. In reality, synchronization was a slowdown in the pace of time, similarly «shrinking» and space. The units of time and space became shorter the faster the speed of the system. But is it a physical reality? Are we hostages of accepted a priori postulates? After all, all these mathematical «phenomena» are nothing more than a consequence of our desire to maintain the constancy of the speed of light, which is not at all convincingly proven by the Michelson-Morley experiment. But what if the experiment proved something completely different, for example, the impossibility of using Descartes' coordinate system for calculations? However, physicists have established the constancy of light intensity as a law. But why, in this case, not to connect time itself with the standard of motion – the speed of light in the void. Analyzing in detail the physical side of clock synchronization in various systems, Einstein, however, is silent about how to measure the distance between them, given that these distances are constantly changing, because systems are in motion relative to each other.

And why the speed of movement of one physical process is a law, and the speed of movement of all the others is not? Apparently, this does not correspond to the real physical world.

Through our mathematical matrices and laws, we look at the world as its reflection in a distorted mirror.

So, in the course of our analysis of the main stages in the development of the views of physicists on nature and the concept of time, a path of more than 2000 years was covered, from Aristotle, as a «number of motion», to the time of Minkowski, as a «fiction». A long path in time, about time itself, logically closed in a philosophical loop. That the «number of motion», that the «fiction» is one and the same, in relation to the real physical world. The nature of time as an independent physical reality has not been revealed. Yes, this could not be done. As the analysis of concrete historical facts and physical theories shows, the concept of time is closely connected with the development of human civilization. It was introduced first into the practice and everyday life of people, and then into science as a reference movement, i.e. some abstraction. But at the same time, it was understood a priori that the physical nature of time and motion is one and the same. However, movement can exist without time, and time without movement does not exist.

THE CONCEPT OF VELOCITY IN PHYSICS

It was shown above that time, as a physical reality, does not exist, and what we call time is nothing but the standard of uniform motion that we have chosen. Consequently, we compare all movements with each other in relation to any one movement. In practice, there is a rationing of all movements, and then their comparison with each other. Normalization shows how many times more or less the speed of movement of the observed bodies in relation to the movement of the reference body (clock).

For the first time, the definition of what the speed of a body is was given by Autolycus of Pitama in the 4th century BC:

«A point is said to move uniformly if it travels equal and equal magnitudes in equal times».

And further:

«If any point, moving uniformly along any line, passes uniformly two segments, then the ratio of one time to another time, during which this point passes one and the other of these segments, will be the same as the ratio of one segment to another segment».

Attention was drawn to two features:
– The ratio of the same veliin is always taken – time and distance.
– Movement speeds are compared, either on the basis of distances traveled at the same time, or on the basis of the time intervals taken to cover the same distance. (3).

It can be seen that this is a purely experimental approach for determining the speed (velocity) of movement of bodies. Moreover, the very concept of «speed» does not yet appear in the calculations. Apparently, this is due to the fact that ancient scientists during this period could not yet connect various physical quantities measured in different units in a single formula. Their mathematical thought has not yet reached the level of relations between two quantities. Describing the outside world purely experimentally, proceeding primarily from the needs of practical activity, they always operated with one category in comparison with another.

The modern concept of speed, as well as the classification of movements and its graphical representation, appeared much later, already in the 14th century, thanks to the works of Albert of Saxony and Nikola Oresmus. (6). What motivated them on this path? Perhaps, only one thing is to find a mathematical equivalent of the speed of movement or a characteristic by which it would be possible to compare all movements with each other. Naturally, this could not be done in the definition of Autolycus. He gave a purely experimental interpretation of the speed of movement, which could not be «squeezed» into the framework of the formula, although it is quite satisfactory for comparing movements, but after measurements.

Let us consider what actually happens in space during the uniform motion of two bodies.

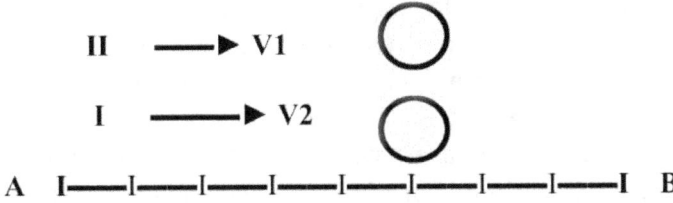

Picture 1.

Points **A** and **B** are fixed relative to each other, Fig.1. Let the bodies have velocities **V1** and **V2** and move in the direction **A-B**. At point **A** there is an observer, which we choose as the origin. What can an observer say about the speed of movement of these bodies, if he does not know anything about the clock? In all likelihood, he will not be able to say anything and determine the relative speed of their movement if he is only at point **A**.

To compare the movements of these two bodies, a second observer at point **B** is required, or the first observer, having made a count of the beginning of the movement, must quickly move to point **B** before the body reaches it and take a second count of the end of the movement. Indeed, if the observer is only at point **A**, then how can he fix, say, the simultaneity of the arrival of bodies at point **B**, if **V1=V2**. If he will be only at point **B**, then how will he be able to determine the beginning of their movement. If we have two observers, at points **A** and **B**, then they will be able to estimate the speed of movement of bodies relative to each other, comparing the results of observations at these points. However, it will be possible to compare the results only if one of the observers moves to another point. Or, as already mentioned, the first observer, from the starting point of the movement, will have to move to another point – the final point.

Then he will have the results of two measurements and will be able to compare the relative speed of the movement of bodies – «who is faster» and on what segment of the path. And although, a priori, it was assumed that points **A** and **B** are fixed relative to each other and represent some kind of abstraction, but as soon as an observer appears in this system, it ceases to be a purely mathematical structure and acquires the status of an inertial system. The observer must have real reference bodies or some surface, for example, the earth. Therefore, in our example, there are two inertial systems **A** and **B** with observers in space. In general, although these systems are motionless relative to each other, they are in complex motion in the space of our solar system and galaxy. It is clear that the observer is an integral part of the process of measuring movements. Thus, we inevitably come to the conclusion that it is impossible to make any measurements in the inertial frame without being in it.

What gives us the introduction of a reference uniform motion – time?

For clarity, let's combine the movement of time and bodies in one figure. This is possible for the following reasons. The observer, having a clock, in general, is in some kind of complex movement in space. Then, every second, a fixed mark will appear in space, caused by the clock cycle. Let's call these marks the points of time **TV**, and the points passed by the body of the unit of the path – the points of the space **TP**. Moving along the same direction **A-B**, the trajectory of the first body will look like a straight line with **TP** marks applied on it. Similarly, the movement of the second body will be a set of **TP**-marks. Let us denote the trajectory of the first body **A1B1**, and the second body **A2B2**. The movement of time in the direction **A-B** will also be depicted as a straight line, with **TV** marks applied on it. At the same time, it is clearly seen that the speed of time, for describing the speed of movement of bodies, does not play any role. In space, we got a picture of **TP** and **TV** marks, Fig.2. If, in reality, in nature, the bodies always arrived at the **TP** marks simultaneously, as in the figure, then the **TV** marks would simply be superfluous. With any combination of **TP** and **TV** marks, we would always get the same value of the speed of movement.

```
A1  I——I——I——I——I——I——I——I——I  B1
A2  I——I——I——I——I——I——I——I——I  B2
T1  I—I—I—I—I—I—I—I—I—I  T2
```

Picture 2.

Each occurrence of a **TV** mark would coincide with the appearance of a **TP** mark. But, since this is not observed and the movement of real bodies is diverse, the need naturally arises for **TV** marks as reference ones. However, the picture is further complicated by the fact that an observer must be introduced into this system. And this, in turn, will require to determine some bodies of the beginning and end of the reference.

Let, further, **V1** is not equal to **V2** and, for definiteness, we assume that **V1=0.5V2**. Then for the same time, **T1-T2**, the first body will cover only half of the path **A2B2**, fig. 3. But the observer, being at point **B2**, will fix only the second body and its time. If the observer is at point **B1**, then the second body will «fly» past him, and the ion will also not be able to fix the arrival of the body at point **B2**.

```
A1   I——/——/——/——I  B1
A2   I——/——/——/——/——/——/——/——I  B2
T1   I—I—I—I—I   T2
T'1  I—I—I—I—I—I—I—I   T'2
```

Picture 3.

In fact, in space, we got two pairs of segments of **TP** and **TV** marks. In order for the observer to be able to fix both bodies, it is necessary that the end of each of the segments **B1** and **B2** coincide with the observer, who, in our example, is at point **B2**. The observer, for a time equal to **T1-T2**, will be able to fix only the second body. In order for it to be able to fix the first body, it is necessary to continue the movement of the first body and time, i.e. counting **TV** marks to the value of **T'1-T'2**. Therefore, the observer, in practice, in principle, will not be able to compare these two movements with each other and determine some characteristic of the speed of their movement. People are accustomed in their lives to determine the characteristics of the movement of various modes of transport, according to already predetermined measurement data. Similarly, this also happens in space during the movement of rockets and planets.

The way out of this situation, just, and give hours. They are always located at the point where the observer is located, and their joint movement in space is rigidly connected. This is equivalent to the fact that the length of the **T1-T2** segment is always «stretched», counting the **TV** marks until the first body arrives at the observer's location. Thus, in practice, the observer is always placed in the center of reference of any movement. He is only interested in the time spent on the road or the distance to the meeting point, using already existing points of reference. And he does not think about how all this looks in an empty space when there are no tools for measurements.

The development of mathematics and the introduction of standard measurement measures made it possible to introduce the ratio of two dissimilar quantities into practice and science. This made it possible, to compare the speed of movement of bodies with each other, to introduce the following combination into physics:

$$V1=A1B1/T1T2 \qquad V2=A2B2/T'1T'2$$

which is called the speed of the body. In the future, always used and these ratios were compared. The unit of distance or length **A-B** is formally different from the unit of time, and the method of their measurement is also different. But what underlies these values? They are defined through movement. In this case, the speed, by its physical nature, is a dimensionless quantity and is simply a multiple of two quantities. It is possible to obtain other combinations of these two pairs of segments, provided that they are of the same physical nature:

a) **(A1 B1) (T1 T2)** and **(A2 B2) (T'1 T'2)**
b) **A1 B1 – T1 T2** and **A2 B2 – T'1 T'2**
c) **A1 B1 + T1 T2** and **A2 B2 + T'1 T'2**
d) **T1 T2** and **T'1 T'2** with **A1 B1 = A2 B2**, etc.

It can be seen that the last combination is the definition of the speed of movement of the bodies of Autolycus, provided that the bodies move uniformly. In principle, all these combinations can be used to determine the speed. It is interesting that the word speed itself comes from the adjective fast, from which they originated: st. slav. – *Speed*. And this applied, first of all, to the characteristics of the person himself. Also, a synonym for speed is fast, which comes from the old. slav. – *bystr* «sharp, fast, lively».

Given that the nature of the movement of real bodies and time were considered different, you can immediately discard the combinations "**b**" and "**c**". It would be impossible to subtract or add two completely different realities measured by different units. This is equivalent to the fact that the length would have to be subtracted from the weight. The combination "**a**" is not suitable due to the fact that when changing **AA** or **TT** by the same amount, the speed would have to increase. All of these combinations required some preconditions. Experience showed something quite different. Namely, if the length of the path is increased proportionally by some value, with the same speed of uniform movement, then the time also increases in the same proportion. There was only one way out – to take the ratio of these two quantities as a characteristic of the speed of movement and introduce a new unit of measurement – speed. At the same time, the absolute value of the speed, in the case of uniform motion, remained constant. It is important that in this case proportional changes in the length of the path and time did not affect the change in the velocity.

Such a combination turned out to be very convenient for practical calculations, and they began to use it everywhere, without thinking about how it is possible to take and divide «meter» to «second». In reality, we divide two numbers and get their ratio, and to this ratio we attribute a fictitious unit of measure **m/s** (meter per second). Thus, from the characteristics of the person himself and his activity, through the development of mathematics and experimental measurements on the surface of the Earth, there was a fundamental revolution in physics, in describing the movement of bodies. Time acquired the status of reality, like space, and organically entered the formula of speed.

Let's pay attention to one significant point. When determining the fact of the arrival of the body at point **B**, the observer must have a clock and two signals, about the beginning of the movement and its end, in order to determine the speed. In this case, a priori, the length of the path **A-B** is considered to be known in advance. But this will require two participants in the observations. The task of the observer at point **A** is to give this signal, according to which the countdown of time will be started and the movement of bodies will begin. He does not need to have a clock at all and, therefore, there is no question of clock synchronization. In this case, the bodies, starting at the same time to move according to the signal, but with different speeds, come to point **B**, at different times, which should be recorded twice by the observer at this point.

At the same time, such concepts as an observer, a point **A** or **B**, a clock are used very freely. It turns out a kind of mixture of abstraction and real life. What about empty space? Let we have two bodies, two observers and a clock. Will they be able to make their measurements? Apparently not. They do not have reference points, there is neither the beginning nor the end of the path of movement of bodies. An observer at point **A** can give a signal for the movement of bodies. But for this, the bodies must first accelerate to a certain speed, and only then move uniformly from point **A** to point **B**. It is not possible to «catch» two bodies in space to start sending a signal for measurements. Also, the second observer at point **A** will record two time readings. But he does not know the distance and cannot determine how the bodies moved. The task becomes generally impossible if there is only one observer with a clock in space. In order to bring some certainty, he must connect his stay with one of the bodies. But in this case, this body becomes a fixed center of reference for the observer.

It will record that the second body is moving past it and even turn the clock on and off at some point. But what will be measured? Some time interval will be measured, on an unknown section of the path. The observer is unable to determine the path of the second body. Moreover, he observes only the relative motion of the second body, but he does not know how and with what speed his own body is moving, as a reference point. Therefore, it is fundamentally impossible to determine the absolute velocity of the second body. Let us leave aside for the time being such a speculative construction of SRT as the arrangement of rods and clocks in space. (10). In any case, in reality in space, we do not observe this.

DYNAMICS OF RELATIVE MOTION OF BODIES

Thus, having established above that time, in its physical essence, does not represent any real quantity different from the movement, it is natural to ask the question: *«Does it make sense to keep it in the equations of mechanics and use it to compare the speed of movement of bodies with each other?»*

If not, then how to distinguish between movements of a different nature: uniform and uniformly accelerated, variable, circular, what to put in correspondence with them?

Let's consider a simple example for clarity. (Fig. 4). We ourselves are an observer from the outside and can mentally imitate any situation. There is a trajectory of the body with initial **A1** and final **B1** points. For measurements, there are standards of length and time. Then on the segment of the path **A1-B1**, the observer will fix the time on the clock **T'1-T'2**.

```
A1  I——————I——————————I——————————I————————I  B1
T'1 I—I—I—I—I—I—I—I—I—I—I—I—I—I—I—I  T'2
A2  I——————I——————————I——————————I————————I  B2
T1  I———I———I——I——I———I———I————I—I  T2
```

Picture 4.

And already on these measurements, he will calculate the value of the speed.

However, the standards of length and time were chosen based on the real conditions of people's lives. We can change the length standard and get a different value for the **A-B** path segment. In reality, the length of the path will remain the same **A1B1=A2B2**. Quite a different picture for measuring time, which is associated with the hands of the clock. By changing the standard of time, the watch will show a different time, on the same segment of the path. In our example, the observer will record a smaller value of

33

time on the hands of the clock. Consequently, the calculation of the speed will also change in the direction of its increase. But, if we assert that time is the same reality as space, then the observer must receive the same speed of the body, regardless of the chosen standards.

We will proceed, first of all, from the following real premises. Our world is space and matter, which is in perpetual motion and interaction. This is a world outside our consciousness and being. It moves, develops and changes according to some of its own, yet unknown to us, laws. And in this world, or rather, in its very small area, an observer appears who wants to know these laws and describe mathematically this world in relation to himself. Where does he start his research? He will cognize the whole external world through his observations and comparisons of the movements of some matter, which in his ideas looks like the movement of real material bodies. And already through these movements or trajectories, the observer will try to determine the laws of their force interaction and some characteristics, for example, the mass of bodies. But let's think about his statement: «*I observe the movement of a body in space*». In reality, this means that light or some other signals constantly come from the observed body through space to it. And the observer fixes already these signals. It is important for him to know the magnitude and the law of change in the distance from him to the observed body. The movement of other bodies relative to each other, he also learns through a change in these distances and their further comparison. The whole process of observation and comparison occurs only relative to the observer. He is not able to move from one body to another and make measurements directly between the bodies that interest him.

Uniform movement in a straight line implies the absence of interaction of material bodies. This is an empty, «dead world», devoid of reality. However, for theoretical reasoning, we the right to assume that in his small area of space, the observer can fix the «quasi-uniform» movement.

On this basis, consider the following example. Let we have two limited areas of space A and B. (Fig. 5).

In area A there are two bodies **I** and **II**, which move uniformly with the same speeds **V1** and **V2**. We cannot consider the motion of only one body, because the presence of an observer always implies the presence of another body, with respect to which the process of motion should be studied.

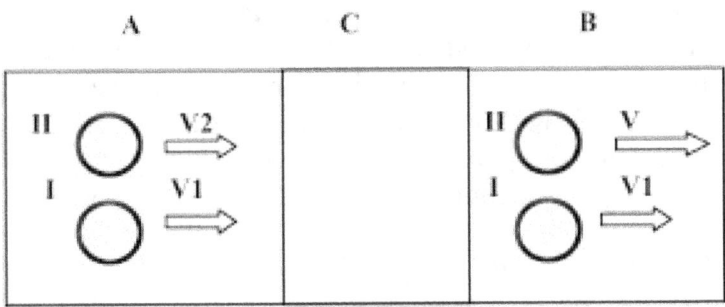

Picture 5.

In area A there are two bodies **I** and **II**, which move uniformly with the same speeds **V1** and **V2**. We cannot consider the motion of one body only because the presence of an observer always presupposes the presence of another body, with respect to which the process of motion should be studied. This is the center with which man is connected and which sees only one body. We will assume that the observer is connected with the body **I**. Initially, the observer will be interested in the movement of the body **II**, which is located nearby and moves at the same speed. Long-term observations of the body **II**, over time, will interest the observer less and less, and then he will completely stop his observations. He will consider that this is the law of his world, everything should remain so forever.

Let at some moment, in the hidden region **C**, an accelerating force act on the body **II** in such a way that the observer did not notice it. Soon, he will again pay attention to body **II**, because will fix that the distance between him and the body is changing. Actually, this will happen in some other region of space **B**, where body **I** will still have speed **V1**, and body **II** will have speed **V2>V1.**

35

The observer will determine that the distance between him and body **II** increases evenly. But, since its observations are carried out from the side, the farther the body **II**, the slower this distance will change. His interest will again weaken over time, since the second body will generally leave his field of vision and he will not observe it. With all these observations, he cannot say anything about his movement, i.e. about the movement of body **I**. Thus, the observer will objectively fix the relative movement of body **II** to the right, with some speed. However, if the observer, at this moment, is combined with body **II**, then he will also fix the relative movement of body **I**, with the same speed, only to the left.

However, the situation is changing. Being in the hidden region C, the observer will feel the impulse of the accelerating force, thereby he will receive a signal that «something has happened». First of all, he will pay attention to the fact that the body **II** begins to move relative to its position of the body I, the mutual distance changes. Knowing nothing about the nature of the force interaction, he may well draw conclusions, as in the first case, when the force was applied to the body **II**, with the only difference that its own speed has not changed, but the speed of the second body $V1$ has changed $>V2$. Only having data on both observations, he will have to introduce some differences for them. He will notice that in the second variant the body **II** moves relative to it not to the right, as before, but to the left.

The most difficult situation for him will be when both bodies receive the same acceleration, while they acquire the same speed of uniform motion. In this case, the observer will also have a signal that «something has happened», he will feel this accelerating force impulse, but he will not be able to observe anything. Body **II** will still be next to him. Not knowing about the first two observations, he will conclude that nothing happened, and such «shocks» are normal for his world.

Based on this example, two characteristic features can be distinguished:

a) to describe the world and build its model, a relative increase in information is needed, which is based on the interaction of matter;

b) to describe the position of two bodies relative to each other, it is enough to know the distance between them and the law of its change.

Under the relativity of the increase in information, one must understand the circumstance that, in general, it must be determined in relation to each of the bodies, if there are no initial conditions. With all observations or laboratory experiments, we just strive to fulfill these two conditions. They are universal, so let's call them «observation conditions».

Let us consider various variants of the motion of two bodies relative to each other, using these observation conditions, in the usual coordinate system. Let two bodies **A** and **B** move in this **XOY** system along parallel trajectories, with the same speeds (graph 2). This movement, from the point of view of our conditions, is of no interest to the observer. It can be said with confidence that throughout this movement of bodies, it will not lead to any interaction between them. For simplicity of analysis, we exclude their interaction through the gravitational field. Then the condition for the distance between them **S=Const** will always be fulfilled.

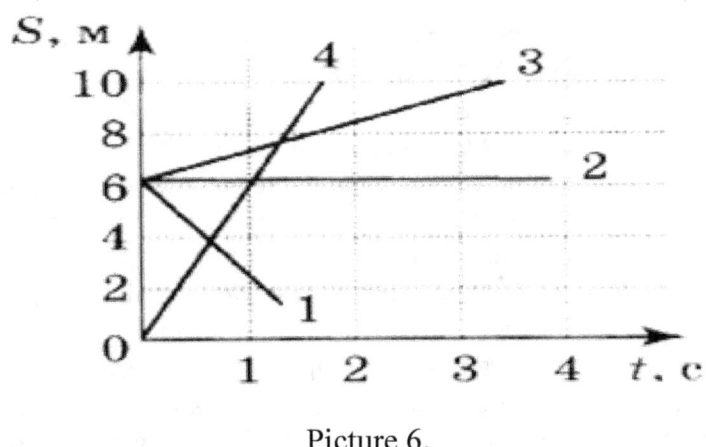

Picture 6.

Now let body **B** move at an angle to body **A**. This option is already of interest to the observer, both from the point of view of the second condition – changing the distance between the bodies, and from the point of view of the first condition – the possible interaction of bodies (their mutual impact). Bodies, having, in general, different speeds, can meet at some point in space, or they can pass by, at some distance from each other (Fig. 6) graphs 3,4. Of all the options shown in the figure, it is the cases of intersection of the trajectories of two bodies that are of the greatest interest.

At these points, the interaction of two bodies occurs and it is possible to study them, i.e. establishing some laws. Indeed, in practice we always strive to learn the interaction of matter in any of its manifestations. Consequently, the third condition of observation will be the requirement – the tendency of the distance between the bodies to zero $S \longrightarrow 0$. It can be seen that all three conditions of observation are closely interconnected. From the increase in information, when the distance between the bodies changes, to their interaction, i.e. $S=0$. Further, this process is repeated again, through the growth of information, to a new interaction. A certain chain is obtained, which determines the process of cognition of nature through its observation.

If time is not a physical reality, then a legitimate question arises:

«What comparative characteristic for the speed of movement of bodies is it advisable to have?»

The expression for the speed of a body, which is now used in physics, is nothing more than an abstract or mathematical ratio of the values of two segments in space. It shows how many times the segment of the path is more or less than the segment of time. Moreover, it is emphasized that we are talking about a period of time, and not a measurement of time, in the sense of seconds and hours. Further, these ratios are compared for different bodies and, thereby, it is determined which of the bodies moves faster or slower than the other and by how much. Denoting the segment of the path through **S**, and the time through t, we can write for two bodies:

$$(S1: t) - (S2: t) = k \text{ or } n1 - n2 = k$$

But what does the value "**n**" or "**k**" mean, in the sense of the physical reality of the movement of bodies. These quantities are faceless, and therefore nothing more than mathematical abstractions. Moreover, the relation, as a mathematical expression, leads in some cases to a nonlinearity or uncertainty of the result, which introduces serious difficulties in building an adequate model of the world. For the movement of bodies, as well as for space, the value of extension is characteristic. Knowing that time marks and space marks are extended segments of the same physical reality – movement, we can determine the magnitude of the speed of the body's movement, relative to the movement of time, as:

$$(X2 - X1) - (t2 - t1) = f(x) \qquad 1.1$$

If the difference does not change during the motion of the body and $f(x) = Const$, then we say that the motion of the body is uniform. It can be seen that expression 1.1 does not have the disadvantages that are inherent in the modern expression for speed. However, it more objectively describes the speed of the movement of the body, relative to the movement of time and has the property of extension. Denote accordingly:

$$X2 - X1 = S; \quad t2 - t1 = T;$$

Then, taking into account the errors in measurements, lengths of time and path, we will have for uniform motion:

$$S \pm \Delta S = T \pm \Delta T + Const$$

From here:

$$S - T \pm (\Delta S - \Delta T) = L \pm \Delta L = Const$$

The definition of the difference **L**, by which we identify the type of movement, can only be made with a certain error. Only up to this threshold $\pm \Delta L$ it can be argued that the motion of the body is uniform. At the same time, the smaller the error, the more insignificant force interactions the observer will be able to fix. This is, first of all, interesting for experiment and measurements.

However, we defined the expression for comparing the relative speed of the movement of bodies using the concept of time, i.e. reference movement. In reality, in space for the observer, all movements are equivalent, he is not able to single out the reference movement. Since the nature of time is the same, the observer must be able to determine and compare the speed of movement of bodies, regardless of whether he has a clock or not.

Suppose that we observe in space, along the O-X axis, the uniform motion of two bodies A and B with different velocities V1<V2.

We have the ability to measure the paths they take, i.e. we have coordinates, but we don't have a clock. (Picture 7).

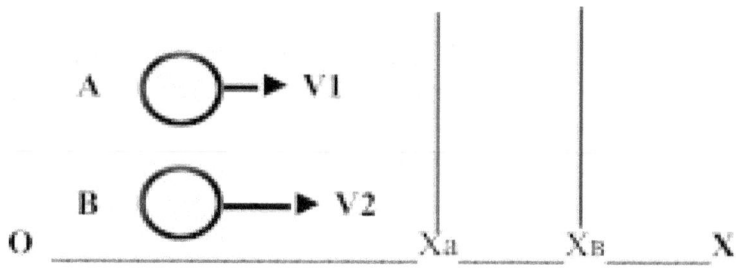

Picture 7.

The fundamental difference here is that there are no timestamps or reference clock movement on the chart. For the objectivity of measurements, we would have to associate the observer with one of the bodies, but it makes no sense to determine the speed of movement of a single body. To do this, we must have at least two bodies and, depending on what interests us, state which body relative to which moves faster or slower. In our mental experiment, we can do this purely qualitatively, a priori, by determining in words that body B moves faster than body A. It is absolutely not necessary to introduce a numerical characteristic for the relative motion of these two bodies, in the sense of their movement speed. The number is introduced with the appearance of the third body. Here it is already necessary to distinguish between the two concepts «faster» or «slower», i.e. how much one is «faster» faster than the other «fast» or «slow». For, as in this case, the observer will have to classify the movements of these three bodies. He will have to say: *«The first body is moving faster than the second body, and the third body is moving faster than the second body, but slower than the first»*.

This is where it is necessary to introduce labels for the concept of «faster», i.e. enter a number and a new characteristic of movement, which today is called speed. We only note once again that for the case of three bodies, when determining the speed through time, in space there will be invisibly present the reference movement of the fourth body – the clock. Moreover, thanks to the clock, which is always associated with the observer, it is possible for him in practice even assign one body a number characterizing the speed of its movement.

Of course, this is provided that the length of the path for him is known in advance. But this is only an apparent picture. Indeed, in reality, in space there is a movement of three bodies: the observer, the clock and the observed body itself. What can the observer in our example determine? It is usually assumed that the observer is stationary and is associated with the **XOY** coordinates. He can only assert the following:

*«When the body **B** is at the point **XB**, then the body **A** will reach the point **XA**»*.

From this it is clear that the movement of body **B** is taken by him as the initial or reference, relative to which the movement of body **A** is determined. Naturally, the body that has traveled the longest segment of the path will have greater mobility. The quantitative measure of this mobility is the difference in the paths that the bodies have traveled until the moment of their measurement, i.e. distance between points **XA** and **XB**. In contrast to the classical concept of speed, where the difference of two ratios is used to compare the movement of two bodies, here the difference of two absolute values is used. Moreover, we are talking about real physical quantities associated with space. Time is not a physical reality, so the use of ratios of two quantities for mathematical calculations clearly conflicts with reality. The use of clocks for experiment is, in principle, not possible. After all, then the magnitude of the relative velocity of bodies would have a different value, if it was determined relative to body **A** or relative to body **B**. Experience, however, tells a different story. The picture of the relative movement of bodies looks the same, both when observed from body **A**, and when observed from body **B**. Therefore, the value that determines the speed of their movement must also be the same.

But what does the question mean for the observer:

«At what moment will body A, moving along its trajectory, reach the point XA, or body B, which is moving in parallel, will reach the point XB?»

And here, being in an empty space, he will not be able to answer anything. After all, observing the movement of several bodies in space, he connects all the moments of their movement with their relative location. Asking such a question, it is necessary to clearly stipulate the first condition, with respect to the motion of which body we want to determine one or another moment in the motion of other bodies of interest to us. In the case of the considered example, we can say that body **A** will move to point **XA** when body **B** is at point **XB**.

Based on our historical experience and the innate habit of watches, one can object: «Well, what does this actually give? If we are traveling between points **A** and **B**, at a speed of 30 km per hour or 60 km per hour, then it is possible to somehow navigate when we arrive at the final point, whether we will be late for the train or not». Indeed, this is so, but only from the point of view of our state and social structure, our labor and everyday needs. Before we look at the clock, many preconditions are imperceptibly fulfilled. We must not forget that this uses the reference movement of the clock, which is synchronized over the entire surface of the earth and all human activities are tied to this clock. Checking the time with his watch, the passenger does not even think about the fact that the train driver also looks at his watch. At the same time, he is well aware of the speed of his train and the distance that is still left to travel to the station. Everything obeys a single rhythm and movement of the clock. For the driver to arrive from point A to point B at the appointed time is the law of his work, and for passengers it is quite natural that the train arrived on time. And, of course, no one will be satisfied with the answer to the question: «When will the train arrive?»

If they hear:

«At the moment when the plane arrives on such and such a flight».

If we have a watch, but no timetable for the arrival of the plane, then we will remain in complete ignorance, we can only wait and guess. Therefore, what is good for everyday life does not at all mean a true reflection of the physical world. The world of man is connected with the surface of the Earth and all mathematical descriptions have been developed and tied to life on this surface. This also applies to the concept of time, speed, coordinate system. It is surprising that, having believed in the infallibility of the laws discovered by him, a person decided to use the same concepts and laws in open space, and even in empty space.

Above, we determined that one of the conditions for observation is the determination of the distance between the bodies and the observer. Consequently, the definition of the speed of movement of bodies will also be determined through these distances. In this case, for a mathematical description, unlike the previous example, we must bind the observer to a single point on the coordinate axis. It can be, first of all, the origin O and consider the movement of the body relative to this point. Consider the uniform motion of the same body with different velocities $V1 > V2$. Let's combine these two movements on one chart. (Fig. 8).

Let the body, moving more quickly with the speed **V1**, move from

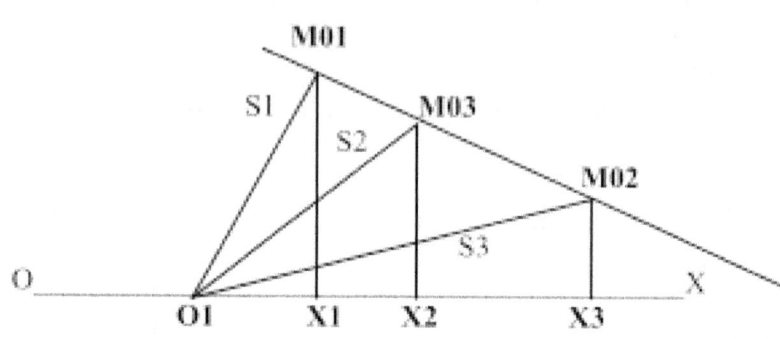

Picture 8.

point **M01** to point **M02**. The same body, moving at a lower speed **V2**, moving from point **M01**, will reach point **M03** at the moment when it, in the first case, reached point **M02**. The beginning of observations and measurements of these movements will be assumed simultaneously from the point **O1**. Then, respectively, for both options, as can be seen from the figure, we can write:

$$\Delta S1 = S3 - S1$$
$$\Delta S2 = S2 - S1$$

The observer will objectively notice that with a faster movement of the body, the amount of change in the distance from it to the body has a relatively large value **S3 > S2**. Therefore, this value can serve as a characteristic of the relative speed of movement of bodies in space. We denote it by **D** and call it the dynamics of the relative motion of bodies.

If in our example, we count the dynamics for the second case, we get:

$$D = \Delta S = \Delta S2 - \Delta S1$$
$$D = S3 - S2$$

44

In the general case, for the motion of many bodies, choosing one of them as the initial level of the speed of motion S0, we get:

$$D1 = S1 - S0 \quad D2 = S2 - S0$$
$$Dn = Sn - S0$$

If it is necessary to make calculations with respect to some other movement of the body, say the first one, then we will accordingly obtain:

$$D2 = S2 - D1 = S2 - S1 - S0 = S2 - \Delta S1$$
$$D3 = S3 - D1 = S3 - S1 - S0 = S3 - \Delta S1$$
$$\cdots\cdots\cdots\cdots\cdots\cdots\cdots\cdots\cdots\cdots\cdots\cdots$$

$$Dn = Sn - D1 = Sn - S1 - S0 = Sn - \Delta S1$$

Since, in all transformations, the value **S0** is present, we can set it equal to zero. From this, the values of the dynamics **Dn** will not change with respect to each other. This just proves the position that in the real world there are no particularly distinguished directions of movement of bodies. Any movement can serve as a reference point, and it is possible to assign any number to this beginning. From this, the description of the motion of bodies in space in relation to each other will not change. Giving the definition of the dynamics of the body, we can say the following. The dynamics of the relative motion of the observed body is the distance by which it leads or lags behind the reference body on a certain segment of the path. It would seem that the bodies should be and move on the same line or along the same trajectory. If their movement occurs along parallel trajectories, then this law is not fulfilled, since the distance to the observed body will always be measured at different angles, and, consequently, the magnitude of the value of the dynamics of their relative movement will be different. But this paradox is nothing more than the result of using the Cartesian coordinate system.

This system was also developed by Descartes for the conditions of applying calculations on the surface of the earth and does not at all reflect the observations of the movement of bodies in open space, which are objectively described by a sphere. A priori, we can say that also the Descartes coordinate system is not universal and cannot be used to describe the relative motion of bodies in space.

It follows from the definition of dynamics that if the magnitude of the dynamics does not change when the body moves in space, then the movement of the body is considered uniform. If it changes on some segments of the path, then the movement of the body on these segments of the path is uneven. At the same time, it is impossible for the observer to determine, without having the initial conditions, which of the bodies was affected by the force. Through measurements, he will record a change in the relative dynamics of the movement of bodies, but it will be impossible for him to determine which of the bodies received an impulse of force action. He will be able to determine this only having a relative increase in information on both bodies for their force interaction. But this does not make his description of the world any better. Of course, using the concept of time, we build our descriptions on the illusion that our clocks run absolutely evenly, like some kind of reality in all space. In any case, we believe and strive for it. And even the slowdown of the clock in SRT, when moving from one coordinate system to another, does not affect the uniformity of their movement. But the real world is an arena of power interactions, both in small and large. This means that the description of the world, using the concept of time, is just as unreasonable. Moreover, it does not reflect the physical realities of the ongoing processes in space. Let us write down all variants for the dynamics of the relative motion of the body:

a) uniform movement:

$$D = ST - S0 = Const$$

b) uniformly accelerated motion of the observed body

$$D = (ST \pm \Delta ST) - S0 = ST - S0 \pm \Delta ST = DT \pm \Delta DT$$

c) uniformly accelerated motion of the reference body:

$$D = ST - (S0 \pm \Delta S0) = ST - \Delta S0 \pm So = DT \pm \Delta D0$$

d) uniformly accelerated motion of both bodies:

$$D = (ST \pm \Delta ST) - (S0 \pm \Delta S0) = DT \pm (\Delta ST - \Delta S0) =$$
$$= DT \pm \Delta DT0$$

It can be seen that for the general case of calculating the dynamics, the formula completely coincides with the expression for describing the movement of the observed body relative to the clock, as their particular movement in space. Therefore, the concept of dynamics also objectively characterizes the speed of the relative motion of bodies in space. Determination with its help of differentiation between uniform and accelerated movement is possible up to the level of **±DT0**. If earlier this distinction was determined by the accuracy of the clock and the determination of coordinates, now it is determined by the uniformity of the movement of the selected reference body and the accuracy of determining the distance to the bodies. However, the general formula for the dynamics of the relative motion of bodies describes the real process of their movement in space, since it operates only with distances and does not create difficulties with uncertainty when **t = 0** in calculations. Not knowing about the impact of a force impulse, i.e. without initial conditions, the observer will not be able to distinguish which of the bodies was affected; for his measurements, the bodies continue to move uniformly. When a force impulse is applied to both bodies, the observer will be able to fix only the difference in this impact, as a relative change in dynamics. In practice, this means that the observer will always perceive a change in the dynamics of the reference body through an apparent change in the dynamics of the observed body. Therefore, for mathematical calculations of dynamics, it is convenient to apply formulas a) and c), when determining the uniform and uniformly accelerated movements of bodies.

Within limited limits, it is possible to admit the existence of a certain plane, like the earth's surface. Then using the Cartesian coordinate system and the same notation, our graph will look like in fig. 9.

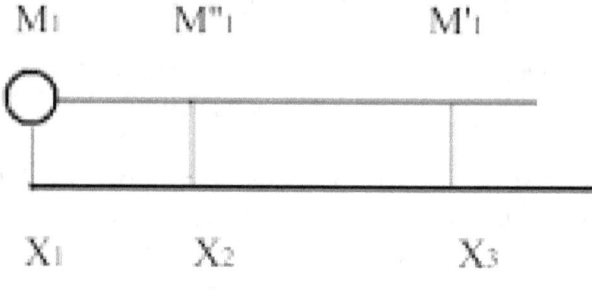

Picture 9.

The origin point **O** coincides with the origin **X1**. Distance readings **S1; S2**; S3 are always equal to each other, so they can be taken equal to zero. In this case, the dynamics of the movement of the body will be determined by the reference points of the distances **X1-X3** and **X1-X2**:

$$D1 = X3 - X1 = XT1$$
$$D2 = X2 - X1 = XT2$$

The dynamics of the relative motion of the body for these two cases is defined as:

$$D12 = XT1 - XT1 = \Delta D12$$

We must, however, remember that we will never determine the magnitude of the absolute effect of force on the body, as well as its absolute dynamics of motion. It is simply unrealistic to impose any initial conditions on the external world. This is permissible only for mathematical equations, in order to obtain some physically justified solutions that, with a certain error, are consistent with the experimental data.

Otherwise, the absolutely uniform motion of the reference body would be silently accepted again, similar to how Newton's mathematical time is now used in all formulas. Consequently, objectively, the observer always determines the relative magnitude of the dynamics of the movement of bodies, and hence the impact of the force.

CONCLUSIONS

First of all, let us dwell once again on the ancient question of the physical nature of time. The answer to it can be given very clearly and unambiguously. Time as an independent physical reality does not exist. It was introduced into practice and science as an artificial reference movement. Having connected time, through the cyclical phenomena of nature, with the hands of a clock, a person believed in its objectivity. Having synchronized all social and labor activities, subordinating everything to a single rhythm and movement, watches have become an indispensable attribute of everyday life. Scientists, using time and calculating the movements of celestial bodies or the flight time of a rocket, remained delighted with the results obtained.

This is where you need to stop!

The human world is the world of the earth's surface. In this world, man has already measured all the distances and made maps. But it is inconvenient to always carry a map with you, and a person made a timetable for the movement of all types of transport and all areas of activity, linking it all to the clock. Each person has his own clock, but he must always compare these clocks with the international standard. What happens if a person finds himself in an empty space? In this case, the clock is rendered useless. There is no time, apart from his own biological clock, a person cannot even know whether he is moving or not.

Is it justified to use the concept of time to describe the real movement of bodies in open space? As our analysis shows, time is a fiction and should disappear from the mathematical equations of physics. Of course, the clock will remain in the practical life of people. But our worldview must fundamentally change. The above conditions for observing the relative motion of bodies show the incompatibility of the physical interpretation of the concept of speed and real motion. Using the concept of the dynamics of the relative motion of the body, it is necessary to make a decisive transition from the space-time world to the world of

49

space and the relative motion and interaction of matter. This will greatly simplify the equations of physics and create an adequate picture of the world.

The clock, as a device that recreates the reference movement, must remain within the earth's surface, as well as many other devices. It is impossible to extend moments of time to all space, they can only be where there is a person himself who is able to create clocks and use them. Therefore, the introduction into physics of such a concept as the dynamics of the relative motion of bodies and the development on its basis of new formulas for the mathematical description of the universe is fully justified.

LITERATURE

P.S. Kudryavtsev "History of Physics" V.1,3. Uchpedgiz, M. 1956

A.V. Butkevich, M.S. Zelikson "Eternal Calendars" Ed. Science, M. 1984

Ya.G. Dorfman "World History of Physics" Ed. Science, M. 1974

V.N. Novosiltsev "On the history of the basic units of the SI" Ed. Rostov University, 1975

S. Michal "Clock" Per. From Czech. Ed. Knowledge, M. 1983

M. Gliozzi "History of Physics" Per. From Italian. Ed. Mir, M. 1970

Aristotle "Physics" Sotsekgiz, M.-L 1934

I. Newton "Mathematical Principles of Natural Philosophy" Op. T. VII. 1936

"Principle of Relativity" Collection. Atomizdat, M. 1973

A. Einstein "The essence of the theory of relativity" Ed. Foreign literature, M. 1955

ANNOTATION

«Physical foundations of the radial-spherical
coordinate systems».

The widely used method of Descartes' coordinates for describing the motion of bodies did not allow eliminating all discrepancies between the calculated and experimental data. This served as the basis for the transition to the curvilinear Gaussian coordinates. However, as the analysis shows, both methods, being formal mathematical constructions, do not make it possible to adequately determine the characteristics of the movement of bodies using experimental data.

The non-linearity and relativity of the process of observing the movement of bodies leads to the need to use such a coordinate system, in which certain physical prerequisites would be laid. It is necessary that the observer be able to describe the external world as he himself sees, observes and measures the movement of bodies.

In this regard, a radial-spherical coordinate system (RSCS) is proposed. This makes it possible to avoid non-linearity in the description of the motion process when moving from one coordinate system to another.

Basic definitions and relations are introduced for finding the distances and coordinates of the movement of bodies, as well as formulas for the transition from one RSSC system to another.

«PHYSICAL FOUNDATIONS OF RADIAL-SPHERICAL COORDINATE SYSTEM»

INTRODUCTION

The study of the real world and the description of the movement of bodies in space are inextricably linked with the concept of a reference system, one of the bases of which is the coordinate system. Borrowed from geometry, it was considered at the beginning as a convenient and illustrative example of a graphical representation of the movement of bodies. In the future, moving on to mathematical equations, no one doubted its validity, having the image of the trajectory of the body in the form of a line. It seemed quite natural that all this appears to us, observing from somewhere from the side. And this is where we run into a contradiction. On the one hand, there are trajectories of movement of real bodies, and, on the other hand, the observer himself is absent in the system. Moreover, the process of physical measurements is completely ignored. But is it possible to assert that these measurements are adequate to the mathematical calculation? Are there discrepancies here or, as a consequence, artificially introduced non-linearity in order to bring them into line? It is possible to obtain a match between the results of calculation and experiment, but they still will not reflect the reality of the observation process.

The historical analysis of the development of the method of coordinates, just allows you to answer these questions. We can imagine the physical world without an observer, but it is impossible to «tear off» the observer from the physical world. He must be present in this world, which he studies and wants to describe. However, at the same time, his mathematical method must comply with the physical principles of observing the movement of bodies in space, and not on the surface of the Earth. Mathematicians introduce the third coordinate into the Descartes system and claim that it is applicable to the calculations of the motion of bodies in space, others introduce the polar coordinate system, and so on. Finally, transformations and calculations of the special theory of relativity (SRT) have come to replace them. But, oddly enough, none of these systems reflects the reality of ongoing events. There are no coordinate systems in empty space, and with uniform motion, the observer sees only linear relationships between the movements of bodies, no matter where he is.

At the same time, his mathematical method must correspond to his physical principles of observation, how he sees and what he sees. Therefore, in this method, the observer, his physical process of perceiving the world and the possibility of displaying the trajectory of the movement of bodies based on measurements should be connected together. Only in this case it is possible to speak about the adequacy of the mathematical model and the real physical process. In this work, just an attempt was made to build a new method and derive new relationships that would meet all the specified requirements. In this case, the results of the previous work will be used, where the concept of the dynamics of the relative motion of the body was introduced and it was shown that the concept of «time» as a physical reality does not exist. (1).

ORIGIN AND DEVELOPMENT
OF THE COORDINATE METHOD

For the first time, in the history of science, graphic representation was introduced by Nicola Orem. (2). In his treatise "On the Configuration of Qualities" (1350), he introduces a geometric interpretation in explaining various concepts of the movement of bodies. By this time, all types of movements had already been classified, the concepts of speed, acceleration and average speed of the body were introduced. (4). The very idea of a geometric interpretation of motion was already beginning to emerge when determining the instantaneous speed, in the case of a differential motion (uneven), when a moving point draws a line. In his work «Quality Intensities», N. Orem proposed to depict this dependence in the form of lines that are drawn from the points of a straight line corresponding to «extensities». Thus, the relationship between them is depicted in the form of a figure, which was limited from below by a direct extensiveness, and from above by some kind of curve. Turning to the kinematics of the movement of the body, he puts the concept of time in correspondence with the extensiveness, and the speed of the movement of the body in accordance with the intensity. (3). From this point on, all kinematic proofs acquire a more geometric character. This was facilitated by both the clarity of the proposed method and the successes of geometry in the future. As you can see, N. Orem does not yet have a second coordinate axis. He used only one axis of «time», from the points of which the perpendiculars. Their length corresponded to the value of the speed of movement body. Perhaps the presence of only one axis was the reason that he did not directly indicate anywhere that the area of the figure in question corresponds to the distance traveled.

So, N. Orem had an image of the trajectory of a moving point, as lines in the plane of a geometric pattern. His main merit was that he made this trajectory dependent on another quantity – time, depicting it as a straight line. Did N. Orem assume in his geometric constructions the presence of an observer, and even more so, his methods of observing and measuring the movement of real bodies in space? Of course not. He used purely formal mathematical constructions and used the concept of speed as a ratio of two quantities. Its coordinates were only as a convenient and visual way of displaying the dependent and independent variables in relation to each other.

Historically, *«...this is something like a transition from coordinates on the earthly or celestial sphere, known in antiquity, to modern coordinate geometry»*. (5).

This is where the separation of the observer and the coordinate system took place. The observer becomes something like a demigod, and the coordinate system becomes his world, in which he establishes his own laws for any distances. At the same time, the observer himself is outside this world and observes it from the side.

N. Orem's treatise «On the Dimension of Forms» was published several times and inevitably had a great influence on Descartes, becoming the foundation of his research. And already here the method of coordinates is fully revealed as a mathematical technique. Descartes' goal was to create a universal mathematics. He defined it this way:

«*...the field of mathematics includes only those sciences in which either order or measure is considered, and it is completely unimportant whether these are numbers, figures, stars, sounds, or anything else in which this measure is found*». (4).

Thus, he intended to create pure mathematics, with the help of which, using pencil and paper, one could describe the whole world. And Descartes went to this mathematics through the unification of algebra and geometry. He put two basic principles at the basis of his philosophy: *relations and extension*. At the same time, extension was understood only as referring to the external figure of bodies and abstracted from everything else. The unit of length can be any measure common to all considered quantities.

It is depicted as a straight line segment and is «the basis of all relations».

Introducing algebraic notation, Descartes develops the method of coordinates not only for displaying dependencies, but also for mathematical calculations. Descartes directly pointed out that he applied his method to the study of nature.

True, in his «Geometry» (1637), there are still no Cartesian axes familiar to us, and, even more so, equations of lines and conic sections. But, by combining algebra and geometry, he significantly expanded the scope of its application and laid the foundations of analytic geometry.

Descartes logically established the relationship between the coordinates on the plane and algebraic equations, being able to represent them as a line. He wrote:

«*All points of lines that can be called geometric, i.e. which fit under some exact and definite measure, are necessarily in some relation to all points of a straight line, which can be expressed by some equation, one and the same for all points of this line*». (5).

In the second part of his Geometry, Descartes extended his method to the three-dimensional case.

Exalting mathematics as the queen of sciences, Descartes absolutized its methods and used it without any restrictions to describe the physical world. The visibility of the coordinate method, the external similarity of the formation of a line during the movement of a point and the trajectory of a body during mechanical movement in space, led to the illusion that it is possible to impose a coordinate grid on the entire external world. At the same time, as before, an important point was excluded from the point of view of the physics of real phenomena. Namely, in the system there is absolutely no observer, his method of observation and measurement. Thus, the coordinate method of Descartes separated theory from experiment, and is a purely formal mathematical construction in the plane of geometric patterns. Can it be argued that the observations and measurements made by the observer in space correspond to the picture that was previously depicted on paper? Apparently, the answer should be negative. After all, all measurements, when a body moves in space, the observer makes relative to himself, i.e. it is located directly in this system, and the model of the coordinate system is a view from the side, in the plane of the picture. Therefore, the inevitable consequence was that, with the accumulation of experimental data, a discrepancy was revealed in the description of that real physical world and those calculated data that were produced using the Descartes coordinate method. They were replaced by Lobachevsky-Riemann geometry and curvilinear coordinates of Gauss. But did this save the situation? Lobachevsky, as a mathematician, only said that he managed to create a consistent geometry, replacing the postulate of the parallelism of two straight lines. The fundamental point in this case was that the geometry begins to be built on surfaces.

«And, most importantly for us, Beltram showed that in ordinary (Euclidean) space there are surfaces that, in this sense of the word, carry the two-dimensional geometry of Lobachevsky». (6).

56

Later, Riemann formalized this method by introducing the concept of «manifold». In his formulation, any surface can be considered as a two-dimensional manifold or a set of points. The set of straight lines in space is a three-dimensional set of points, and the set of spheres is already represented by a set of four dimensions, etc. How should an observer feel in such a geometry? He would have to observe the movement of bodies also on these surfaces. But where in the open, empty space should arise some complex surfaces, moreover, of many dimensions? How could they even be located there, on what sides of this very surface? But after all, space has the property of isotropy. How then to explain the change in its geometry, which must inevitably lead to a change in the movement of bodies.

The presence of gravity of other massive bodies that cause these changes in motion in some local areas of space, how is this explained in the general theory of relativity? But at the same time, changes in the movement of bodies are determined by the magnitude of the force interaction of two or more bodies. Do we have the right to attribute these changes in the movement of the observed body to the geometry of space itself?

The curvature of the trajectory of the observed body is the result of the interaction of gravitational fields, i.e. force interaction of matter itself, and not a property of space as such. Moreover, it has been proved by the methods of astronomical observations that, in any case, up to a value of the order of 10^{23} km, no curvature of world space is observed.

Nevertheless, the existing experimental data confirm the closeness of the results to the calculations based on the methods of non-Euclidean geometry. This fact, apparently, is explained by the following. The method of the Cartesian coordinate system does not allow one to obtain mathematical models for describing the world that are adequate to the real physical world. Indeed, this system is quite consistent with our practical experience and the calculations that are made on the surface of the earth. But with increasing distances, moving to the open space of space and increasing the accuracy of measurements, it leads to non-linearities of our transformations. At the same time, there was only one goal – to link the measurement data with mathematical calculations. It is these inconsistencies, as well as new views of the special theory of relativity (SRT), that have been attempted to be compensated by the introduction of non-linear transformations during the transition from one coordinate system to another.

Let us analyze this example in more detail from the point of view of the conditions for observing the uniform motion of a body in Descartes' rectangular coordinate system. Based on the first condition, the impossibility of measurements in two inertial frames at once must be assumed that the observer is located at one of the same coordinate points. This is the starting point for real observations and, if possible, real measurements of the movement of the observed body. This is all the more true if we consider regions of space of great extent. The observer is not able to first fix the beginning of the movement, and then transfer it to another point in space, with an almost instantaneous speed, at which it is supposed to fix the end of the measurements and make a new measurement. We connect the observer with the origin of some imaginary coordinate system **O**. In reality, the entire process of observing and measuring the observed body will occur along a straight line connecting the body and the observer **S**, i.e. along the line of sight of the observer. The process of its measurements will also take place along the same straight line. Therefore, in this case, the second observation condition is also satisfied, i.e. the distance from the observer to the body will be measured, throughout its movement, and the law of its change will be determined. In this case, the measurement process itself, in a rectangular coordinate system, will be depicted as in Pic.1.

Let the body, moving uniformly parallel to the **X** axis, move from point **M1** to point **M2**. The observer at the origin of the coordinate

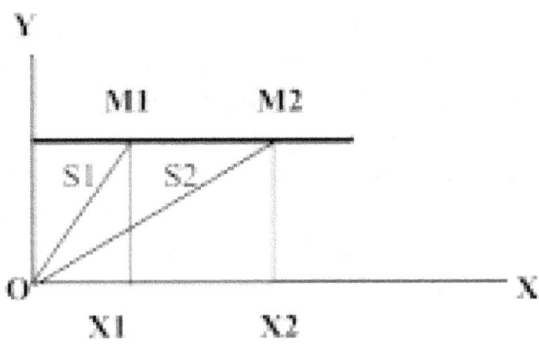

Picture 1.

will fix the distance S equal to:

$$\Delta S = S1 - S2$$

$$\Delta S = \sqrt{(X1 - X2)^2 + Y^2} = \sqrt{(X1 - X2)^2 + b^2}$$

Mathematically, this formula is nothing more than the Pythagorean theorem for calculating the sides of a triangle. But by itself, it only describes a static relationship between the sides of a triangle. This means that there is a calculation of some fixed, immovable position of the body, relative to the observer, or its individual fixed states.

Moving uniformly along the X axis, the body will sequentially pass the coordinates: **X1; X2; X3**, etc. How will the distance **S** change in this case? Taking the derivative with respect to **X**, we obtain the law of change in the distance **S** from the observer to the observed body:

$$dS = \frac{X}{\sqrt{X^2 + в^2}} = \frac{1}{\sqrt{1 + ß^2}}\, dX, \quad \text{где} \quad ß = \frac{1}{X}$$

In practice, this equation determines the transition from the coordinate system, as a mathematical method, to the physical measurements of the observer, i.e. from the measured value **S** to its mathematical equivalent **X**, when the body moves.

It can be seen that this transformation is clearly non-linear. With a uniform rectilinear motion of the body, which is an objective physical reality, the observer in this coordinate system will fix the uneven motion of the body. The nature of its measurements is shown in Fig. 2.

DS/dX

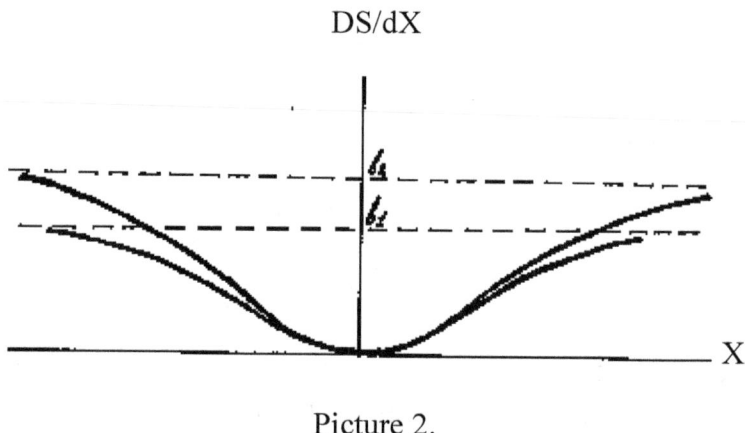

Picture 2.

The observer will inevitably conclude that some force acts on the body, which leads to a change in its dynamics during movement. This discrepancy between the experimental data and the calculation by the method of Cartesian coordinates made it possible to reveal the accumulated experience. What did the curvilinear coordinates bring then? They allowed the **X**-axis to be «curved» in such a way as to compensate for this discrepancy as much as possible. In reality, in such a coordinate system, it is not a description of the movement of the body itself in space that

60

occurs, but the movement of its image along a hypothetical surface. Having brought the data of theory and experiment into conformity, scientists did not receive an adequate model of the movement of the body in space. Indeed, according to the conditions of observation, it would be necessary to place the observer in the center of this surface, which would make the description of the motion of the body even more intricate and *non*-linear.

What is the reason for the non-linearity between the distance **dS** and the coordinates **dX**? There can be only one reason for this – the absence of an observer and his process of observation and measurements in the system. Consequently, it is precisely the spatial non-linearity of observation and measurements that stands between the observer and the outside world.

This feature should be taken into account when constructing a coordinate system, if it is required that the mathematical dependencies and true relationships in the motion of the body be adequately determined. It makes no sense to talk about the trajectory of some body, in some abstract coordinate system. We should only talk about the movement of a given observed body in space, in the perception of the observer, and already on this foundation try to create a mathematical method based on the principles of physical measurements.

RADIAL-SPHERICAL COORDINATE SYSTEM (RSCS)

Having established the fundamental point that the external real world, refracted through the process of observation and measurement, appears before the observer in a completely different form and in other quantitative assessments, it becomes obvious that it is necessary to apply such a method to describe it, which would eliminate this discrepancy.

Let us consider in more detail what happens in space when observing the movement of a body. Let the body move uniformly along the trajectory **A – B** relative to the observer, who is at point O, Fig. 3.

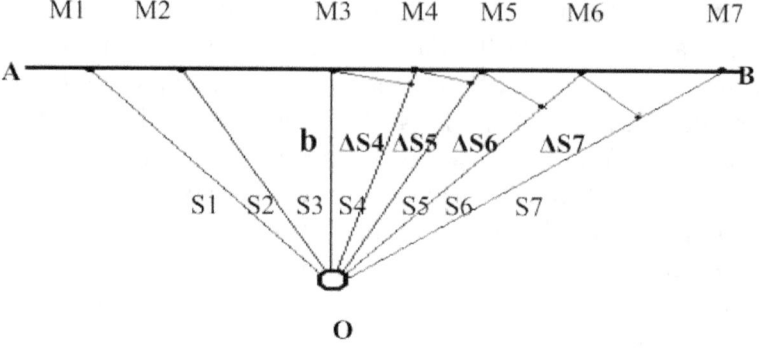

Picture 3.

In order for the observer to determine whether this movement is uniform or not, signals must be sent from the side of the body with the same frequency, i.e. at points **M1M2 M3** ... Mn at equal distances from each other, in the direction of movement of the body.

Only in the event that the observer fixes that the intervals between the arrivals of signals are equal, he can classify the movement of the body as uniform. It is clearly seen from the figure that the intervals between individual signals will be determined by how evenly, i.e. by the same value or not, the distances **S1S2S3 ... Sn** differ. If the condition is met:

$$Sn - Sn\text{-}1 = Sn\text{+}1 - Sn = Const,$$

then the observer will come to the conclusion that the motion is uniform. But what does the observer actually see? It will fix that the signals from the body come unevenly, because. conditions are not met:

$$\Delta S4 \neq \Delta S5 \neq \Delta S6 \neq \Delta S7$$

Thus, the observer will come to a false conclusion about the non-uniform motion of the body. Since the dynamics of the relative motion of the body is also determined through the distance with respect to the observer, which is the reference point, then, consequently, the magnitude of the dynamics will change in the same proportion. Moreover, from Fig. 2 it can be seen that the magnitude of the dynamics will have a different nature of change, depending on the parameter "**b**".

The observer, being in an open space, observes and measures the movement of the body always in a straight line of sight, when the body itself moves at a certain distance from it. This leads to measurement distortions. Under what condition is it possible that the magnitude of the dynamics of the body remains always constant and corresponds to its true value of uniform motion of the body? Obviously, this is possible only if the parameter "**b**" is equal to zero:

$$\mathbf{b} = \mathbf{0}$$

In reality, this means that the trajectories of movement of bodies in such a system must always coincide with the X axis, along which measurements are made. Since the space is isotropic and there are no privileged directions in it, and, moreover, the directions of motion of real bodies are diverse, it is necessary to fulfill the condition that any straight line passing through the point O, where the observer is located, would have this property, i.e. e. could serve as the axis of the coordinate system.

Then for any direction the equalities will be valid:

$$\mathbf{dS} = \mathbf{dX} \quad (2.1) \qquad\qquad \Delta \mathbf{S} = \mathbf{D0} \quad 2.2$$

Only under these conditions, the observer will be able to fix and determine the uniform motion of the body, indeed, as uniform. And, further, already with respect to this body, determine the possible unevenness in the movement of other bodies and build some models of force interaction.

From 2.1 and 2.2 it becomes clear what the coordinate system should be if we want to display adequately the movement of bodies in space relative to the observer. It is necessary to imagine well that the observer is in open space. Moving his gaze in different directions, and fixing it always at some one distance, he will see not straight lines, but a sphere. In the center of this sphere is the observer himself with his instruments, which is the beginning of the coordinate system. Thus, we introduce the concept of a radial-spherical coordinate system (RSCS). Let us agree to call the point O the center of the coordinate system. (Fig. 4).

Concentric circles are drawn from the center of the coordinate system, the radii of which differ per unit length. These circles create a scale on any of the coordinate axes. Physically, the coordinate axis is a specific rectilinear trajectory of the body. This determines the choice of this axis for the observed body from many others. We call these trajectories radial.

Let us analyze an example of uniform motion in the RSSC of three bodies **M1 M2 M3**, which move uniformly and rectilinearly along their radial trajectories: **AB, AB'', AB'''** towards the center O of our figure. For the body of reference we choose the movement of the body along the trajectory **A-B**. We assume that the dynamics of bodies **M2 M3** are the same with respect to **M1**. Let's see if it is possible to determine this at the observer's place, at point **O**, using this coordinate method. For convenience, we will denote the coordinates of points through **Xn**, where the index n is equal to the numerical value of the coordinate. We fix the position of the bodies when **M1** has passed path $S1 = X5 - X4$. In this case, **M2** and **M3** will cover, respectively, the distances equal to:

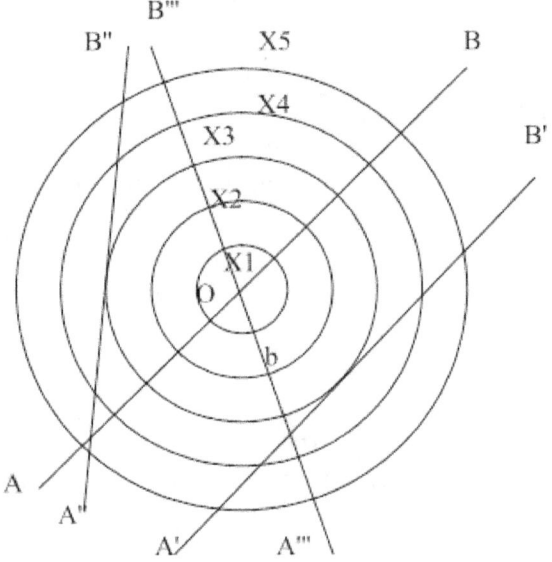

Picture 4.

$$S2 = X5 - X3 = 2$$
$$S3 = X5 - X3 = 2$$

Therefore, their dynamics of relative motion is defined as:

$$D2 = (X5 - X3) - (X5 - X4)$$
$$D3 = (X5 - X3) - (X5 - X4)$$
$$D2 = D3$$

It can be seen that the paths traversed by the bodies **M2** and **M3** in the RSSC turned out to be equal in size, despite the fact that they visually differ. Consequently, the values of their dynamics will also be equal. Since, in other similar areas, the observer will fix the same values of **S** and **D**, he will come to the conclusion about the uniform motion of the bodies. Thus, using the RSSC method, we have thereby established an adequate connection between the description of the movement of the body and the measurement data from the observer, and, on the other hand, the real movement of bodies in space.

If we accept, in the case of a uniform motion of the body, that we have determined the dynamics of the relative motion of the observed body on a single segment of the path of our reference body, then with a radial trajectory, its path will be determined in relation to the observer as:

$$S = S0 + nD0$$

where:

S – the path traveled by the reference body,
n – is the number of single paths throughout the movement of the body.

With respect to the reference body itself, the length of the path is defined as:

$$S = nD0$$

It is obvious that these formulas are valid for any uniform motion of a body at any position of its trajectory on the RSSC graph. Substituting the value of the dynamics for the uniformly accelerated motion of the body (1), similarly we get:

$$S = n (D + \Delta D)$$

How, in this case, will the spatial model of the RSSC look like? Obviously, in this case, instead of concentric circles, there will be spheres of the same radius. The observer himself will be in the center of these spheres. This world of his should be called spherical, which corresponds to his real principles of observation and measurement. The spatial symmetry of the sphere and the requirement of equivalence of all axes in the RSSC are identified with such a fundamental property of space as isotropy. This also makes it possible to establish the linearity of all transformations in determining the relative motion of bodies, while avoiding the non-linearity of the very process of observation and measurement. It is easy to make sure that all the relations of classical mechanics are valid in the RSCS when bodies move in space. The difference lies in the fundamental rejection of the concept of time, as having no real physical basis. In the case of RSSC, the concept of the dynamics of the relative motion of bodies is used, and then the length of the path for uniform motion is determined as:

$$X = X0 + nD0$$

At present, in mechanics, the concept of speed is used and, accordingly:

$$X = X0 + Vt$$

It can be objected that the interaction of bodies is impossible in the RSSC, since the conditions of radial motion are imposed. Events are possible only in the center. But, as was shown earlier, the observer cannot record events at two points in space at once. (1). Since, first of all, it is events that interest us, this means that the observer must be at the place of the event or combine the center of his RSSC with this event. Mathematically, this means to make a transition from one RSSC system to another.

Expressions 2.1 and 2.2, as applied to RCCS, allow us to draw several more important conclusions. Firstly, only in the case of radial trajectories of the movement of bodies, the absolute values of the measurements coincide with the true values of the characteristics of the movement of the body. Secondly, in relation to the observer, all directions of movement of bodies in space are equivalent. It is important for him to know only the distance to the observed body and the law of its change. It does not matter which trajectory the body is moving and in which direction. The meaning of the direction of movement of the body matters only in everyday life. This is due to the fact that all purposeful human activity takes place in the «flat» world of the earth's surface, which is limited to certain points. A person must first arrive at some point, and then return back. Even if he does not come back, then there remains a relative relationship, the starting point and the final point. In contrast to this, the concept of the dynamics of the relative motion of a body is not a vector quantity, like the concept of velocity in classical physics. It characterizes the movement of the body unambiguously and regardless of which side the observer is looking at.

In reality, in space, the trajectories of motion of bodies do not coincide with the radial directions of the RSSC. Is it possible to make any calculations in these cases? Let two bodies move uniformly in the RSSC along parallel trajectories **A-B** and **A'-B'**, with the same dynamics **D**, fig. 4. The trajectory **A-B** passes through the center of the system, and the trajectory **A'-B'** is offset from the center by a value **b**. Naturally, the observer, being in the center and making measurements, must also make the conclusion that the motion of the bodies is uniform and with the same magnitude of dynamics. It can be seen from the figure that the conclusions of the observer will be objective only if the signals from the bodies are sent at the same coordinates **X3 X4 X5** ... etc. However, something else is also obvious. The body moving along the trajectory **A-B**, passes equal segments of the path: **X5 – X4 = X4 – X3**, etc. The body, which moves along a different trajectory **A'-B'**, passes simultaneously different segments of the path, from the moment of sending one signal to another. What is it connected with? Only with the position of the observer relative to the observed body by the value of the parameter b. This is where the effect of non-linearity of observations lies. But does the position of the trajectory **A'-B'** and the direction of motion of the second body relative to **A-B** have any significance for the observer? Obviously not. After all, the second body, for example, can move along the trajectory **A"-B"**. In this case, all mutual relations of movements **A-B** and **A"-B"** and the results of their measurements will be analogous, as in the case of **A-B** relative to **A'-B'**. Interestingly, on the other hand, if we consider the movement of the body along **A"-B"** relative to the movement of the body along **A'-B'**, then all the arguments remain valid, with the only difference that the length of the paths traversed by the bodies will be the same: **X"5 – X"4=X'5 – X'4** etc. Since the observed signals are always recorded synchronously, this will lead the observer to the conclusion that the dynamics of the bodies are equal. It follows that for any trajectory of the body, the distances between two adjacent coordinates are always equal to the same value:

$$L = Xn - Xn\text{-}1 = Const \qquad 2.3$$

The nonequivalence of the absolute length of the segments in the figure cannot be identified with their absolute value in space along the direction of the trajectory itself. This is caused only by the non-linearity of the very process of observation and measurement, which finds its real reflection on the graph.

And, since a person is assigned the role of an observer in space, his observation is limited to a sphere. He must always keep this in mind and not "notice" the unevenness of the segments of different trajectories of the movement of bodies on the RCCS chart. He must get used to the fact that the path distance between two points of RSCS coordinates is always equal to the selected unit of measurement.

CONVERSION OF COORDINATES IN RSSK

On the basis of relation 2.3, we have thereby equalized the observers whose system centers occupy different positions relative to the same observed body. But, having equalized them,

However, it is impossible to equate the results of their physical measurements in this case, in the sense of the RSSC method used to describe the motions of bodies. Each of the observers will fix different values of the change in distances, determined by the parameter – **b**.

It all depends on how far the trajectory of the body's motion passes from the center of the RSSC. Therefore, those coordinate values that are determined for radial trajectories and correspond to physical measurements will be called absolute values, and all the rest relative. In order to find out the relative values of the coordinates through absolute ones and, thereby, determine the true characteristics of the movement of the body in space, the observ-

69

er must make the transition to another RSSC, for which the given trajectory of the observed body will be radial.

Let us explain what has been said with the following example. Let there be two radial-spherical systems RSSK1 and RSSK2 with centers **O1** and **O2**, fig. 5. Consider the uniform motion of two bodies with the same dynamics along the trajectories **A-B** and **A'-B'**, in the case of centers that are motionless relative to each other. In this case, the trajectory **A-B** is radial for both systems, and the trajectory **A'-B'** is radial only for RSSC2. What conclusion will observers come to, who are each in their own center of the system and not knowing about the ratio 2.3?

An observer from the center **O1**, on the basis of his measurements, will conclude that the movement of the body along **A'-B'** in relation to **A-B** is uneven and occurs with different dynamics.

For an observer from the center **O2**, these movements are completely equivalent, and he will naturally come to the conclusion that the bodies move uniformly and with the same dynamics.

A

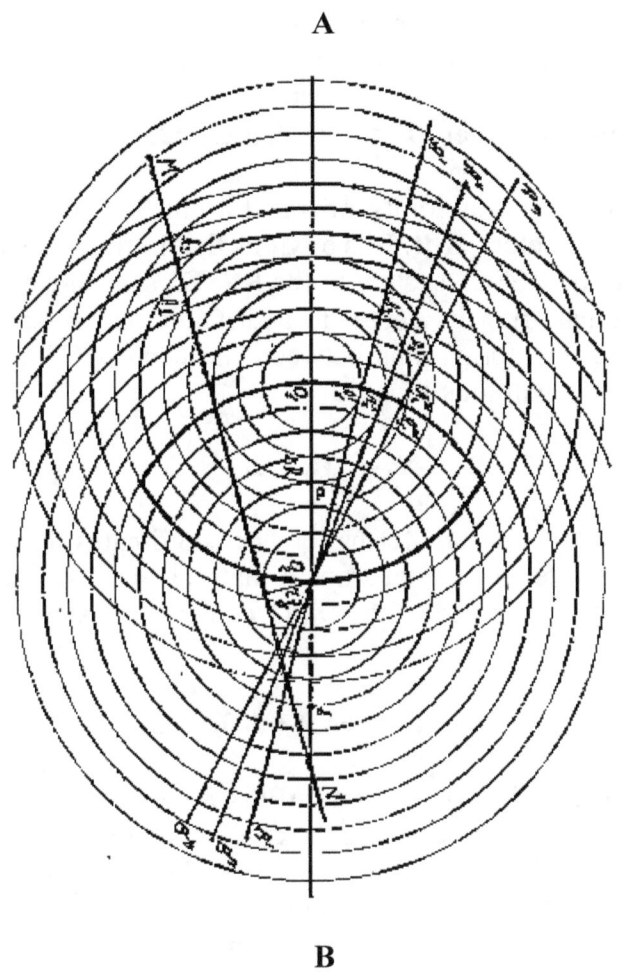

B

Picture 5.

Thus, it is clear that the various conclusions of the observers are due to the positions of their centers of the system: **O1** and **O2**, relative to the trajectory of the bodies. Equation 2.3 allows eliminating these inconsistencies when determining the nature of the movement. However, the quantitative results of the observers, i.e. measurement results will be different. It can be seen from the figure that the absolute values of the distances for an observer in **O2** are not equal to the relative distances for an observer in **O1**, when measured along the trajectory **A'-B'**. Since the true values during measurements correspond only to radial trajectories, the observer at the center **O1**, in order to find out these values, must move from the center **O1** to the center **O2**, i.e. to transform the coordinates.

Consider the possibility of transforming coordinates first for the radial trajectory **A-B**. It is obvious that in this case, the transition from the RSSC1 system to the RSSC2 system can be carried out by transferring its center to a distance $\Delta R = X8$. Then the coordinate transformations between the centers **O1** and **O2** for both systems will look like:

$$X' = \Delta R - Xn \qquad 3.1$$

Let's call these transformations of coordinates for the centers of RSSC systems internal transformations. Then the coordinate transformations outside the RSSC centers or external transformations will be:

$$X'' = \pm\Delta R + Xn \qquad 3.2$$

Where the sign in front of ΔR is determined depending on which side of the center **O2** is the coordinate **Xn**.

Is it possible to determine the same relations in the case of the trajectory **A'-B'**? After all, many trajectories can pass through the center **O2**, which in the RSCK1 system will have the same value of the **X** coordinate. For example, in the figure, the trajectories of movement of other bodies also have the same **X4** coordinate: **A''-B''**, **A'''-B'''**, **M-N**, etc. But in the RSSC2 system, the same point with the **X4** coordinate already corresponds to different absolute values of the measurements. This situation corresponds to reality. For an observer in the center **O1**, points **A2'**, **A2''**, **A2'''** are equal in relation to each other, i.e. are at the same distance from the center. From the side of the observer in the center of **O2**, these distances turn out to be different, which means that the values when transforming the coordinates will also differ.

In this case, if the reference is made not relative to the distance between the centers ΔR, but relative to the sphere of radius $R0 = \Delta R$, then similarly as for cases 3.1 and 3.2, we can write:

for the trajectory A'-B': $\mathbf{XA'2} = \mathbf{\Delta R} + \mathbf{A'1\ A'2} = \mathbf{\Delta R} + \mathbf{X'n}$
for the trajectory A''-B'': $\mathbf{XA''2} = \mathbf{\Delta R} + \mathbf{A''1\ A''2} = \mathbf{\Delta R} + \mathbf{X''n}$

where $\mathbf{X'n}$ and $\mathbf{X''n}$ are already relative values.

We define a sphere of radius $\mathbf{R0}$, respectively, as a null sphere. Thus, we can say that the value of the same coordinate $\mathbf{Xn}$ in RSCK1 is ambiguously determined in RSCK2, depending on the trajectory of the body. We have two observation points $\mathbf{O1}$ and $\mathbf{O2}$, but even the introduction of a third «objective» observer will not lead to an unambiguous choice, since his own measurements will give their coordinates of the observed bodies, different from the first two. Consequently, our process of observation, and therefore measurements, is not only non-linear, but also relative. When the observer moves from one point in space to another, the entire observed picture of the relative motion and relative position of the bodies not only shifts, in the sense of linear distances, but also rotates.

Let's determine the relationship between the relative and absolute values of the $\mathbf{X}$ coordinate. To do this, consider the movement of two bodies in the RSCS along the trajectories $\mathbf{A\text{-}B}$ and $\mathbf{A'\text{-}B'}$, fig. 4. Let the body, moving along $\mathbf{A\text{-}B}$, pass a symmetrical path relative to the center of the system from $\mathbf{X5}$ to $\mathbf{X5}$. Similarly, the second body, moving along the trajectory A'-B', passes the same path with coordinates $\mathbf{X'5}$ and $\mathbf{X'5}$. It can be seen from the graph that the coordinates of the trajectory $\mathbf{A\text{-}B}$, based on relation 2.3, correspond to all coordinates on the trajectory of the body along $\mathbf{A'\text{-}B'}$, with the exception of the area bounded by a circle of radius Rb. Indeed, the second body does not approach the center $\mathbf{O}$, closer than $\mathbf{Rb}$, which means that its coordinates in the RSCS system less than $\mathbf{X3}$ do not exist. Thus, when finding the absolute values of coordinates through relative ones, there is a ban on the transformation of individual coordinates, depending on the location of the trajectory of the body from the center of the system. In our example, it is limited by a sphere of radius $\mathbf{Rb} = \mathbf{X3}$, which we define as a sphere of prohibition. In the general case, the radius of the sphere of prohibition in the RSSC is determined by the smaller distance that will be fixed when the body passes by the center of the system.

Therefore, we can write the following expression for finding the absolute values of coordinates in terms of relative ones:

$$\mathbf{Xabs = Xrel + Rb} \qquad 3.3$$

At the same time, one should always remember that $\mathbf{Xrel \geq Rb}$. Returning to fig. 5, it is then possible for $\mathbf{X5}$ to determine its value in PCCK2 through its absolute value in PCCK1. For the trajectory $\mathbf{A'\text{-}B'}$ we have:

$$\mathbf{XA'2 = \Delta R + (X'n + R'b)}$$

Accordingly, for another trajectory $\mathbf{A\text{-}B}$ there will be a different value of the parameter $\mathbf{b}$:

$$\mathbf{XA''2 = \Delta R + (X''n + R''b)}$$

Thus, for any trajectories of motion of bodies in the RSCS, the following expressions will be valid, respectively, for internal and external transformations:

$$\mathbf{X'n = \Delta R - (Xn + Rb)} \qquad 3.4$$
$$\mathbf{X''n = \pm\Delta R + (Xn + Rb)} \qquad 3.5$$

In this case, however, the size of the sphere of prohibition will already be determined not by the smallest distance of the body trajectory from the center $\mathbf{O1}$, as in 3.3, but by another value. In that case, it was fair as the ratio was determined in the same RSSC system. In this case, there are two RSSK systems and, when transforming coordinates, the value of $\mathbf{Rb}$ should already be determined by the value of the coordinate in RSSK1, when the zero-sphere RSSK2 and the body trajectory intersect. Indeed, if RSCK2 $\mathbf{Xn = \Delta R}$, i.e. is equal to the value of the null-sphere, then in RSCK1 it is necessary that $\mathbf{Xn = Rb}$. It is obvious that this will always be at the intersection of the null sphere of RSCK2 and the prohibition sphere of RSCK1. If the trajectories of motion of bodies for both systems are radial, then **Rb = 0** and expressions 3.4 and 3.5 coincide with 3.2 and 3.3.

In the general case, if the trajectory of the body **M-N** is not radial for any of the centers, then the radius of the sphere-prohibition and the value of **ΔR** can be determined from 3.3:

$$\Delta Rabs = \Delta Rrel + rb$$

Since rb is defined in the same coordinate system, then its value will correspond to the smallest distance from the center of **O2**. Substituting this value into 3.4 and 3.5, we get, respectively, for internal and external transformations:

$$X'n = (\Delta R + rb) - (Xn + Rb) \qquad 3.6$$
$$X''n = \pm(\Delta R + rb) + (Xn + Rb) \qquad 3.7$$

These expressions make it possible to perform coordinate transformations already for any trajectories of motion of bodies in the RSCS. Thus, the application of the RSSC method makes it possible to exclude the nonlinearity of the process of observations and measurement of the parameters of the movement of bodies from the side of the observer, and to present all the relationships in a linear form.

Consider, further, the transformation of coordinates with a uniform relative displacement of the centers of the RSSC with some dynamics **D**. Obviously, in this case, the distance **ΔR** between the centers of the systems will be determined by the expression:

$$R(D) = \Delta R + nD0$$

Substituting this expression into formulas 3.6 and 3.7, we get:

$$X'n = [R(D) + rb] - (Xn + Rb) \qquad 3.8$$
$$X''n = \pm[R(D) + rb] + (Xn + Rb) \qquad 3.9$$

Since, for the RSSC, $\mathbf{X = nD0}$, we similarly have:

$$\mathbf{Dabs = Dorel + Rb}$$

Then, if $\mathbf{\Delta R = 0}$, then we have:

$$\mathbf{R(D) = D + Rb}$$

From here:

$$\mathbf{X'n = (D + Rb) - (Xn + Rb)}$$
$$\mathbf{X'n = \pm(D + Rb) + (Xn + Rb)}$$

In the case of a radial trajectory:

$$\mathbf{X'n = Xn + D}$$

From where the rule for adding the dynamics of the relative motion of bodies in the RSSC system directly follows:

$$\mathbf{D'n = Dn + D} \qquad 3.10$$

It can be seen that 3.10 corresponds to Galileo's velocity addition rule for a rectangular coordinate system.

DETERMINATION OF DISTANCES IN RSSK

To consider this issue, let us turn again to Fig. 5. For radial trajectories of motion of any body, the distance between two points is defined as:

$$S = \sum_{n=1}^{n1}(Xn - Xn\text{-}1) + \sum_{n=2}^{n2}(Xn - Xn\text{-}2) = Xn1 + Xn2 \qquad 4.1$$

In the same expression, the first sum is the path that the body will take, moving along the trajectory **A-B**, from the origin to the center of the system. The second sum is the path that the body will take moving from the center of the RSSC system to the end point. If the trajectory of the body's path is symmetrical with respect to the center of the RSSC system, then $n1 = n2 = n$

$$S = 2\sum_{n-1}^{n}(Xn - Xn\text{-}1) = 2Xn$$

In the case of non-radial trajectories **M-N**, due to condition 3.1, we can write:

$$S' = \sum_{n=1}^{n1}(X'n - Xn\text{-}1) + \sum_{n=1}^{n2}(Xn - Xn\text{-}1) = X'n1 + X'n2 \qquad 4.2$$

where, respectively, the first sum is the path that the body will take to the sphere of prohibition:

$$r_b = x2$$

and the second sum is the path that the body will take from the sphere of prohibition to the end point. From 4.1, 4.2, 3.3 it follows that for any position of the trajectory of the body, the following relation is true:

$$S = S' + Rb \qquad 4.3$$

where **S** is the distance between two points with radial trajectories;

S' – is the distance between the same points of the trajectory, which are separated from the center of the system by the size of the prohibition sphere.

As noted earlier, it is impossible, in the general case, to transform for all coordinates when passing from the RSSC2 system to the RSSC1 system, outside the prohibition sphere. They simply do not exist there, which means that all the distances between the points of the trajectory of the body's movement also vanish.

Again, in the case of a symmetrical trajectory of motion of bodies in the RSSC, we have:

$$TS = 2(X'n + Rb) \qquad 4.4$$

A seemingly paradoxical situation arises. Let us write, formally, the beginning and end of the movement of the body along the trajectory, as $a = X6$; $b = X6$. Outwardly, according to this record, it can be argued that since the initial and final points of the trajectory of the body's movement coincide, then the body is at rest and the path it has traveled is equal to zero. But this comes from the fact that in this case the concepts of a rectangular coordinate system are habitually transferred to the results in the RSCS system. After all, in the case of the given example, different points were determined on the coordinate axis in the RSSC, which were given by the trajectory of the body.

The equality of their values only indicates that they are absolutely equal in relation to the observer. It makes no sense to talk about the trajectory of some body, in some abstract coordinate system. We should only talk about the movement of a given observed body in space, in the perception of the observer, and already on this foundation try to create a mathematical method based on the principles of physical measurements.

Let us consider an example when, during the movement of two bodies, they passed the same distances between two points **A** and **B** along the trajectory **I; II; III**, fig. 6. It can be seen that, due to relation 4.3, the bodies, in relation to the observer, go through different paths, although in space these paths are the same. What is the reason for this discrepancy? When determining the coordinates, the observer always makes measurements according to the signals that he receives from the bodies. Then, in the case of trajectory I, the difference between the signals about the beginning of the movement of the body and its end will be maximum. When moving away from the center of the system, the signals for **II** and **III**, about the beginning of the movement, will pass, respectively, the path **OA'** and **OA''**, and the signals about the end of the movement of the bodies, the path **OB'** and **OB''**.

And, the more this removal of the trajectories of movement of bodies from the center of the system, the closer the values of these distances. Consequently, the smaller the difference between the arrivals of signals from both bodies and the smaller the distance between points **A** and **B** will be determined by the observer.

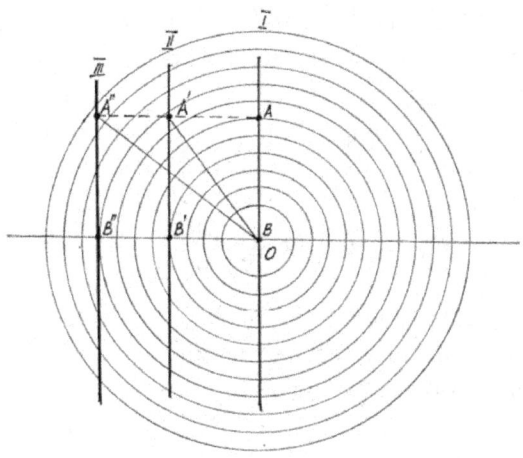

Picture 6.

According to the measurement results, this will correspond to the fact that the distance between the observed body and the observer changes by a smaller amount. With a significant removal of the body from the center, the observer will not be able to determine at all whether the body is moving or at rest, for relatively small displacements. That is why distant stars appear to us to be motionless.

This circumstance made it possible to associate the axes of a rectangular coordinate system with them. However, as we have seen, their immobility is not an objective reality, but a consequence of our perception of the surrounding world. Thus, the method of the RSSC system makes it possible to correlate the spatial characteristics of the motion of the body with the methods of measuring the observer.

In order to determine the true value of the path length, **AB** or **A'B'**, the observer, having the data of his measurements, must transform the coordinate system. In this case, however, one peculiarity arises. We are already accustomed to the fact that if any coordinate is determined when a body moves in one rectangular system and a point is set, then it will certainly coincide with the point that we put in another system if we count its value, as converted coordinate. Having performed the same operation in the RSSC system, it is easy to verify that this correspondence is not obtained. Indeed, let us define, for the general case, on the trajectory of the motion of the body **MN** in the PCSC1 system, the point **A = X6**, Fig. 5.

$$\mathbf{XA2'' = (\Delta R + rb) + (XA' + Rb) = (6 + 2) + (2 + 4) = 14}$$

This will correspond to point A2. It can be seen that these two points do not coincide. What caused it? Only the relativity of the measurement process.

Using relations 3.1, 3.8, and 3.9, we equalize the observers, both in the sense of an adequate observation process and, in the sense of quantitative assessments of the measurement results. Each of the observers, having determined that the movement of bodies is uniform, will attribute to them the corresponding dynamics. Since the values of the dynamics will be equal, then, consequently, the paths traversed by the body in both systems of the RSSC will also be equal. For example, if the body moved along the trajectory **MN** in RSCK1 from the point **X6** to **X4**, then its path will be **S = 2**, respectively. Then in the PCCS2 system for the second observer, we get the same path value:

$$\mathbf{S = (\Delta R + rb + X6) - (\Delta R + rb + X4) = 2}$$

Thus, the task is reduced to being able to recreate the trajectory of the body in a graphic image based on the results of measurements made in another RSCS system.

To talk about the transition from one coordinate system to another, as about some formal mathematical operation, is nothing but the same abstraction that has nothing to do with the real process of observing the world. It must always be remembered that in this case, there are two observers in space, who, in general, move relative to each other and at the same time, each of them makes measurements of the movement of the same body, i.e. some physical process. It is important to always know if the values of the measurements of both observers coincide and to be able to use these measurements. To do this, it is necessary either to combine the observers in one of the centers and compare the results of these measurements, or, artificially, to transform the measurement values of one of the observers into the location of the other, i.e. convert them. However, all calculations, when converting coordinates to RSCS, are performed using absolute values. But the geometric construction also requires to equalize their relative values, which are individual for each of the observers, which correspond to measurements from its center of the system. At any point in space, you can build your own system of RSSC, in each of which linear relations 3.8 and 3.9 will be fulfilled. But any of them, when observed from another center, will already be presented as non-linear. Therefore, it is impossible to use the transformed measurement values without transforming the RSSC system itself in order to geometrically combine the same trajectory of the body movement, which two observers see from two different centers.

In general, when measuring distances in various systems, based on the ratio 4.3, one can write for RSSK1 and RSSK2, respectively:

$$S1 = S'1 + 2Rb1$$
$$S1 = S'1 + 2Rb2$$

Since the values **S1, S2** were defined as absolute, then for the same trajectory of the body, the following condition must be satisfied:

$$S'1 + 2Rb1 = S'1 + 2Rb2 \qquad 4.5$$

This expression is nothing more than a rule for the transformation of the RSSC system itself or its transformation in order to be able to use and graphically represent measurement data produced in another system.

Let's construct two systems of RSSC with the centers in **O1** and **O2** fig. 7. Let's draw radial trajectories through the center of the second **O2** system in such a way that they are relative in the PCSC1 system and pass through each sphere of prohibition. Since, for the RSSK2 system, the measurement values will be absolute and correspond to the true measurement values, it is necessary that in the RSSK1 system they correspond to the same quantitative measurement values. To do this, for each trajectory, on the basis of 3.6 and 3.7, given different values of **Xn** in the PCSC2 system, we find the values of **X'n** in the PCSC1 system.

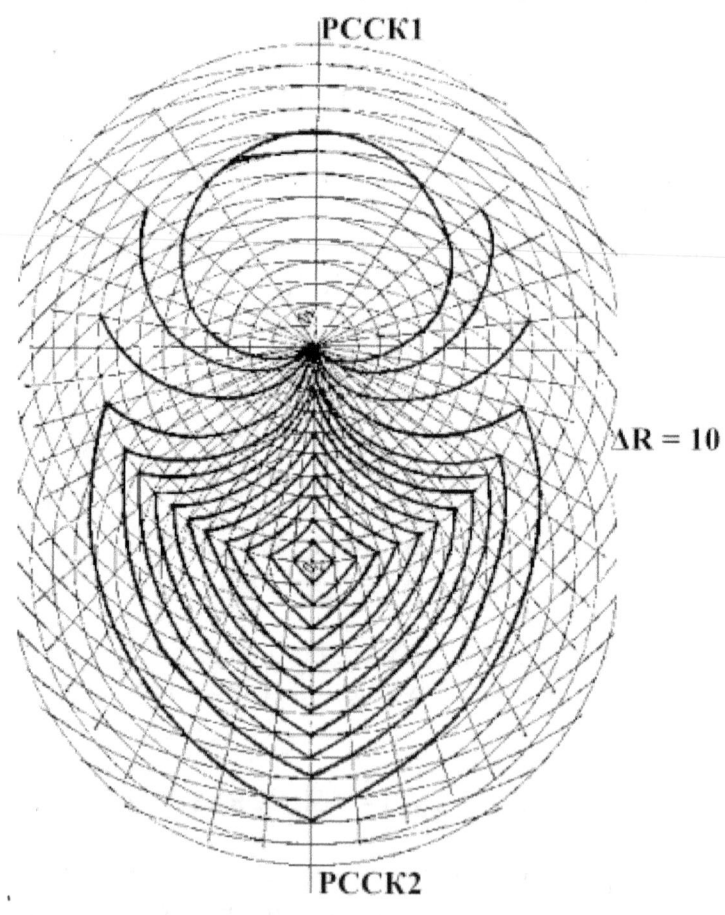

Picture 7.

After that, the corresponding points **X'n** are connected to each other. In this way, we will obtain a transformed system of RSCK1, in which it is already possible to use the measurement data of the system of RSCK2 and which, near the centers **O1** and **O2**, is a kind of «petal». This is how the RSSK1 system will look for an observer from the center **O2** of the RSSK2 system.

Only in this case will coincide any points on the trajectory of the body movement, defined in the RSSC2 system, and then used for calculations in the RSSC1 system. Thus, according to the measurement data, the observer in the RSSC2 system can completely recreate graphically the trajectory of the body in the RSSC1 system. Therefore, if we assert that **X'n** is expressed by such and such a value through **Xn**, then this means that the **X'n** coordinate is already in the transformed system RSSK1, which is defined relative to the center of the PCCK2 system. In practice, it is not at all necessary to build a transformed RSSC system. Simply, you need to remember this. When performing coordinate transformations from the RSCS1 system to the RSCS2 system, we should not look for the coincidence of the coordinates of some of its specific values or the values of **Xn** and **X'n**, in the sense of the coincidence of the intersection of the corresponding RSCS spheres and the given body motion trajectory, i.e. coincidence of the points of the trajectory. Based on relations 3.8 and 3.9, the equivalence of their absolute values is established, when observing a specific body movement from two specific centers **O1** and **O2** of two RSSC systems. Naturally, when observing from another **O3** center, the values of the **Xn** coordinates will differ, but again, they will be equivalent to the first two. Therefore, the construction of a transformed RCCS system is necessary only for a graphical representation of the trajectories of bodies in the RCCS1 system, according to measurement data in the RCCS2 system. At the same time, it will always change depending on the values of **ΔR** and **Rb**. If the movement in the RSSC2 system is not radial, then the «petal» of the transformation of the RSSC1 system will be asymmetric. If the centers **O1** and **O2** of the RSSC1 and RSSC2 systems move in space relative to each other, then the shape and position in space of the «leaf-current» of the RSSC will change.

However, this approach is a purely geometric, formal technique and does not reflect the real essence of physical measurements. Indeed, in fact, by converting the coordinates of the body's trajectory from RSSC2 to RSSC1, we thereby instantly transfer the absolute values of the measurements of the second observer for use by the first.

The second observer, continuing to measure his relative parameters of the same body trajectory, will first bring them to absolute values and only then compare them with the values of the first observer from the RSSC2 system. As already mentioned, these measurement values of both observers will coincide. Thus, from a physical point of view, it will be correct to transform the coordinates in two steps. First, relative values of the coordinates are used to determine their absolute values in RSSK2, and only then, these values are used to find the relative values of coordinates in RSSK1 for the given trajectory. In this case, the values of $\Delta\mathbf{R}$ and Rb are assumed to be known, since the position of the centers of the RSSC systems must be unambiguously determined for specific transformations.

What, in this case, will be an inertial frame of reference, if we use the method of the RSSC system and the concept of the dynamics of the relative motion of the body? Obviously, this will be the RSSC system itself, in the center of which is the observer, with his instruments for measuring distances, and the reference body, which moves relative to the observer uniformly and rectilinearly along any of the trajectories. Let us consider how the dynamics of the relative motion of the body in the RSSC will be determined. At the same time, one must take into account that the center of systems, in the general case, is in motion. This means that having determined and measured the distance to the starting point of the trajectory, the observer will measure the distance to its end point, in fact, from another center. Let the center of the RSSC O1 system move uniformly with some dynamics $\mathbf{D}$, Fig. 8.

We assume conditionally that the reference body moves along the trajectory $\mathbf{A\text{-}B}$, and the observed body moves along the trajectory $\mathbf{C\text{-}D}$. In the initial position of the center $\mathbf{O1}$, the observer will fix the initial points $\mathbf{A'}$, on the trajectory of the reference body, and $\mathbf{C'}$ on the trajectory of the observed body. End points $\mathbf{B'}$ and $\mathbf{D'}$, he fixes from the center $\mathbf{O'1}$. It can be clearly seen from the graph that these will be completely different distances, in relation to those that the observer determined while at the center of **O1**.

We assume conditionally that the reference body moves along the trajectory **A-B**, and the observed body moves along the trajectory **C-D**. In the initial position of the center **O1**, the observer will fix the starting points **A'**, on

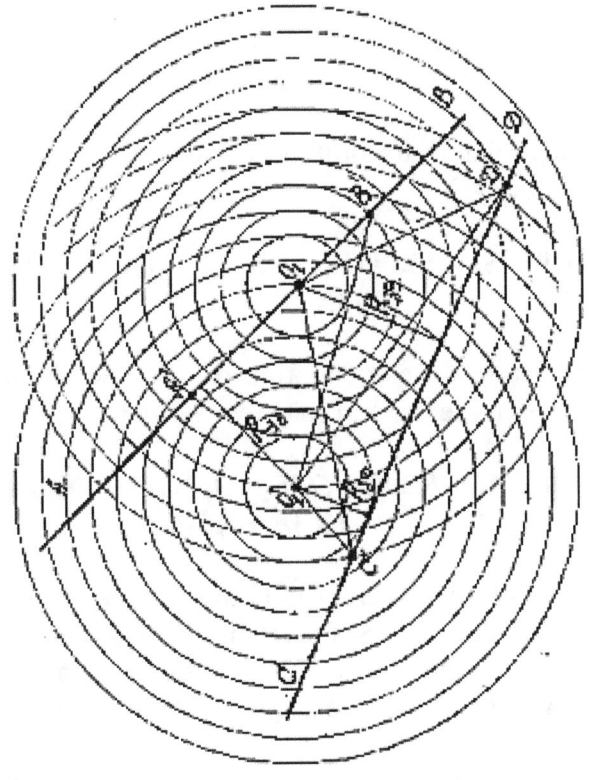

Picture 8.

trajectory of the reference body, and **C'** on the trajectory of the observed body. End points **B'** and **D'**, he fixes from the center **O'1**. It is clearly seen from the graph that these will be completely different distances, in relation to those that the observer determined while at the center of **O1**. Therefore, in order to determine the magnitude of the dynamics of the relative motion of the body, it is necessary to transform the coordinates **B'** and **D'** for the RSSC system with the center in **O1**:

$$Xv' = X'v' + Rzv - D \quad Xd' = X'd' + Rzd - D$$

From here, respectively, the paths traversed by the reference body and the observed body are determined as:

$$S1 = Xa' + X'in' + Rstar - D$$
$$S2 = Xv' + X'd' + Rzd - D$$

Dynamics relative to motion, in the case of radial motion of the reference body:

$$D0 = S1 - S2 = (XA' - XC') + [(X'B' + R3B) - - (X'D' + R3D)]$$

Since the trajectories of motion of reference bodies may not be radial, then, in the general case, we get:

$$D0 = [(X'A' + R3A) - (X'C' + R3C)] + [(X'B' + R3B) - - (X'D' + R3D)]$$

It can be seen from the formula that the dynamics of the motion of the center of the RSSC system does not enter directly into the expression for **D0**. But it has its influence on the process of observation and measurement, which is reflected in the values of **Rz**.

For everyday life, determining the distance between some points using the formulas of the RSSC system is, of course, inconvenient. But after all, they do not use the ratio of a rectangular coordinate system, and, moreover, a curvilinear one. We are always interested in the absolute values of distances, and not in the process of movement between these points, when observed from some centers. Therefore, in practice, it is quite justified from any point, as from the center where the observer is located, to build your RSSK and determine the distances in radial directions. Once the distances to certain points are measured and stored, they no longer depend on the position of the observer. If he is in another place, then he can always use this data. This is also real because a person already knows all the distances on the earth's surface, which is reflected on the maps. He determined these distances in radial directions, constantly moving the center of his RSSK.

CONCLUSIONS

Considering the fundamental points on the way to the development of the coordinate method, you are convinced that it was developed as a purely mathematical construction, and is not adequate to the conditions under which observations and measurements are made in the movement of bodies.

At the first stages, satisfying, first of all, the practical needs of people's activities, when the distances were small, and the world looked flat on the surface of the earth, Descartes' method of rectangular coordinates was not in doubt and became widespread. Its visibility, simplicity and further successes in geometry determined the philosophy of physical thinking.

Further research and discrepancies between experiment and theory led to the need to use curvilinear coordinates to describe physical phenomena. Apparently, this should be considered as a formalized attempt to bring the calculated and experimental data into conformity.

Proceeding from the fact that the process of observation is not only nonlinear, but also relative, it becomes quite obvious that it is impossible to exclude the observer from a system in which certain physical prerequisites would be laid down. The main advantage of the RSCS system is the linear relationship for all trajectories of motion of bodies and all transformations during the transition from one RSCS system to another, both for distances and for the dynamics of the relative motion of bodies. The adequacy of the description and the actual movement of bodies in the observer's perception made it possible to correlate the measurement values with the coordinate values and avoid such artificial conditions as the need to locate the ends of the axes of a rectangular coordinate system on distant stars.

The physical nature of the RSSC method and the linearity of all its relationships allow us to hope that its application in practice and science will make it possible to eliminate those discrepancies between experiment and calculations that are currently present when using the Descartes coordinate system.

LITERATURE

IN AND. Ivashov "Physical foundations of mechanical motion and its description" 1985

M. Gliozzi "History of Physics" Per. From Italian. Ed. Science M. 1970

A.G. Grigoryan, Mechanics from Antiquity to the Present Day. Ed. Science M. 1974

G.P. Matvievsky "Rene Descartes". Ed. Science M. 1976

D.Ya. Stroik, A Brief Outline of the History of Mathematics, Ed. Science M. 1978

V.F. Kagan "Essays on Geometry" Ed. Moscow University of 1963

APPLICATION

MICHELSON-MORLEY EXPERIMENT IN RSSK

The ideas of the special theory of relativity (SRT) originated on the basis of new experimental data, which either contradicted or were poorly explained by existing physical theories and did not fit into the framework of generally accepted fundamental concepts. It is quite natural that at the first stages of comprehending new facts, all efforts were directed towards the modernization of existing theories and concepts in order to preserve the foundations of a legalized worldview. It took a considerable time before the postulates of SRT were formulated and recognized and the concept of the universal ether was abandoned. The Michelson-Morley experiment can rightly be called the decisive experiment on the way to the emergence of the foundations of SRT. It was his «negative result» that shook the theory of the ether, its objective physical existence and pointed to the equality of all inertial reference systems. Einstein himself commented on this experiment:

«The most important of these experiments is that of Michelson and Morley, which I assume to be well known».

Thus, the validity of the special principle of relativity can hardly be in doubt.

Historically, this experiment was considered as a direct proof of the principle of relativity and the constancy of the speed of light. However, in later times, other opinions were expressed about the results of the experiment. In particular, it was noted that his result only proves the relativistic contraction of the length of the body in a moving coordinate system, but did not prove the constancy of the speed of light from the movement of the system. In addition, Michelson himself did not rule out that «the negative result of the experiment may indicate that there is still some kind of ambiguity or incompleteness in the theory itself».

It seems appropriate to dwell on this experiment once again and answer the main question: «What did the Michelson-Morley experiment prove?» This has a huge physical and philosophical aspect for our worldview. Moreover, in our studies it has already been shown that the concept of time is not a physical reality, and the use of Descartes' rectangular coordinate system cannot be applied to the relative motion of bodies in space.

For the purposes of scientific research, it is necessary to use such concepts as the dynamics of the relative motion of the body and the radial-spherical coordinate system. What turned out to be suitable and convenient for everyday life and the purposeful activities of society leads to distortions in the observation process and errors in measurements for scientific purposes.

Let us consider once again briefly, what was the Michelson-Morley experiment? It was based on an interferometer, the arms of which **L1 L2** were located at an angle of 90 ° relative to each other. In the first experiments, the Michelson interferometer had an arm length of 120 cm. 2x10v8 waves of the yellow sodium line were placed on this length. In this case, as follows from their calculations, the expected displacement of the bands was 0.04 of the width of one band. The idea was to use the Earth's velocity V in its orbit to move the interferometer. In this case, when the arms of the interferometer are rotated along the Earth's motion and against it, a shift of the interference fringes should be observed, which could be detected by the available instruments from the center **O**, Fig. 1. Although the interferometer itself was imperfect and almost insensitive to the expected fringe shift effect, Michelson nevertheless concluded in a long series of measurements that no fringe shift occurred. In the future, the increase in the sensitivity of the interferometer occurred along the path of increasing the length of the arms and the use of lasers.

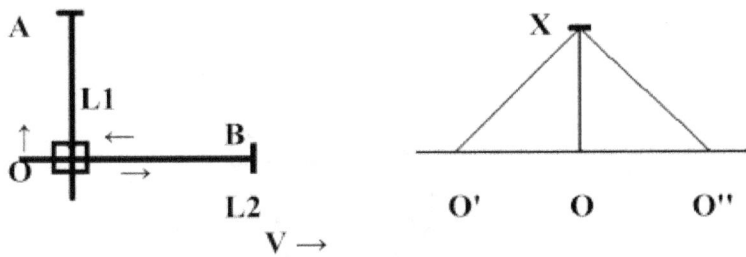

Picture 1.

During measurements, two light signals were sent from the center **O** in the direction of mirrors **A** and **B**. Reflected from the mirrors, the light signals entered the interferometer device, forming fringes of the interference pattern.

In this case, the difference in time and length is defined as:

$$\Delta t = t1 + t2 = (L/c - V) + (L/c + V) = (2L/c)\,(1/1 - \beta^2)$$

$$l = l_0 \sqrt{1 - \beta^2}$$

Naturally, it was expected that the calculated time difference between the arrivals of two signals should inevitably demonstrate a shift in the interference fringes. However, this did not happen!

The theory and calculation of the Michelson experiment was based on the rectangular coordinate system of Descartes, as well as the famous Pythagorean formula: $c^2 = a^2 + b^2$.

Let's consider how the theory of the Michelson-Morley experiment will look like in the RSSC system, using the concept of the dynamics of the relative motion of the body. First of all, let's find out how the Pythagorean theorem will look like in the RSSC.

We have

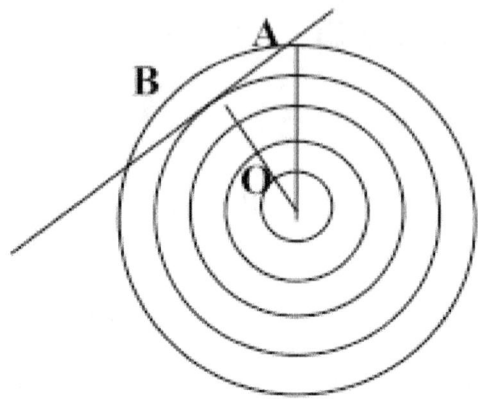

Picture 2.

also a right triangle **OBA**, the aspect ratio of which will be:

AO = OB + BA = 4 +1 = 5

Thus, in the RSSC, the ratio of the sides of a triangle is expressed by a linear relationship.

On the other hand, it is necessary to realize that when setting up this experiment, we are making measurements in one inertial system moving at the speed of the Earth.

No tricks can divide it into mobile and immovable systems, as is often indicated when interpreting the conditions for conducting an experiment. This means that the analysis of the experiment must be carried out exclusively for one moving system.

We will also take into account the fact that the propagation of a light signal does not depend on the movement of the source of this signal. This was proved experimentally, based on the fact that the electromagnetic field is an independent form of the existence of matter. The movement of the source will not bring any changes to the speed of propagation of the light beam. Once, having left the source, it will spread, further, according to its physical laws.

The experiment does not take into account such factors as the Doppler frequency shift or the motion of the Earth following the Sun in the vastness of the galaxy.

The path that the arm of the interferometer L1 will pass along the axis of motion of the earth with dynamics D will look like in Fig. 3:

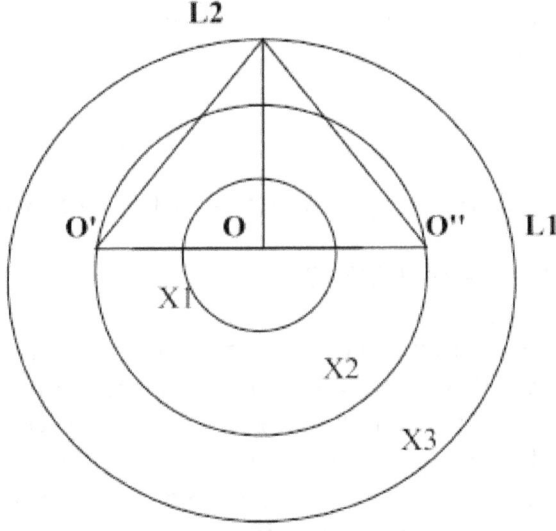

Picture 3.

It can be seen from the figure that the dynamics D will be determined by the coordinates **O-X2**, therefore, the path length of the first light beam, in the direction of the earth's movement, will be equal to:

$$l1 = L + D$$

In reverse direction:

$$l2 = L - D$$

Then the total path will be respectively:

$$S = l1 + l2 = 2L$$

Let us consider carefully what happens to the light signal in perpendicular motion along the arm of the **L2** interferometer. It can be seen that the light beam travels a path

$$X2\text{-}L2\text{-}X2$$

in this case, **L2** is determined by the X3 coordinate. Based on this, it can be written that:

$$S' = 2(L2 + D)$$

Hence, the difference between the paths taken will be:

$$\Delta S = S' - S = 2L + 2D - 2L = 2D$$

When the interferometer is rotated 90°, we will get the same difference ΔS of the paths travelled. Consequently, when the interferometer was rotated, the interference fringes could not shift.

As can be seen, the application of the RSSC system method makes it possible to verify that the Michelson-Morley experiment should not have detected any shift in the interference fringes. The paths traversed by two signals turn out to be equal in space. Michelson's words that there may be ambiguity in the theory itself turned out to be prophetic. During the experiment, it was proved not the constancy of the speed of light, but an error in the application of the rectangular coordinate system of Descartes and the interpretation of the results of the experiment.

In this case, the question is raised about the consistency of SRT and its views on the description of the movement of bodies, on our entire worldview.

Concerning general theory of relativity, then those few effects, such as black holes or redshift, then they can be explained in other, simpler ways, without resorting to such a complex and cumbersome mathematical apparatus. Einstein himself said with regret:

«When mathematicians leaned on the theory of relativity, I myself ceased to understand it».

Let us consider the Michelson-Morley experiment from the other side, namely, from the point of view of the course of physical processes. In the experiment, one light source was used, which was divided into two beams. This raises an interesting question, what are these two beams. In practice, separation is carried out using a translucent glass plate. In this case, each of the rays retains 50% of its intensity, but the frequency and initial phase of the oscillations remain identical. The use of the interference method involves, first of all, a phase shift of two identical radiations. However, the artificial division of one source into two beams does not bring anything new. Similarly, one can use the source itself, the radiation of which is spherical in nature. But, due to the mirrors, two light beams will return to the interferometer, which will create an interference pattern. It is important whether the phase difference will change when the interferometer rotates, fig. 4

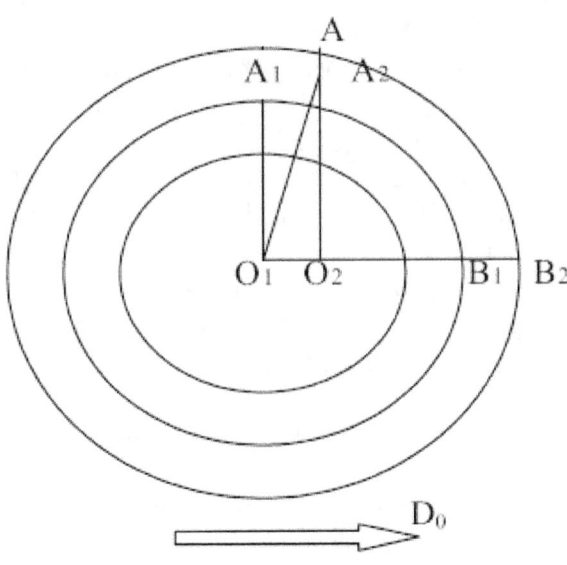

Picture 4.

So, there is a light source **O1**, which emits light waves uniformly in all directions.

The interferometer itself consists of two arms **O1-A1** and **O1-B1**. If the speed of the interferometer is zero, then the phases of the light wave in both directions will coincide: $\delta A1 = \delta B1$. In this case, it should be especially noted that this will be a phase of the same light wave. Reflected from the mirrors **A1; B1**, they again reach the interferometer **O1** and form an interference pattern on its screen. However, in reality, the interferometer is, from the very beginning, in uniform motion with the Earth's orbital velocity **D0**. Therefore, the front of the light wave will reach the mirror at point **B2**, while the same front will reach the mirror at point **A2** earlier. It will already be reflected from it and will move in the opposite direction until it reaches mirror **B2** and is also reflected from it. We have the same front of a light wave, which has a path difference in the opposite direction from two mirrors due to the movement of the Earth, which is physically expressed in the movement of both the mirrors themselves and the interferometer.

In the opposite direction, the light front from mirror **A2** will travel the same path as in the forward direction, i.e. the total path will be double the length of **O1-A1**. However, the light front from the mirror **B2** will travel a much shorter path, since the interferometer moves towards it at a speed **D0**, which eliminates the difference in the passage of paths and it will always be equal to twice the length of the arm of the interferometer **O1-B1**. Thus, there is a phase difference of light waves from two mirrors, which actually represent the same light wave. Due to the separation on the mirrors and their movement in space, different conditions for the propagation of light waves were artificially created. It is clearly seen from the figure tha

$$2[O1-A2] > 2[O1-B1]$$

Their difference explains the phase shift of two light waves. This will be the first interference pattern or the beginning of a possible shift of these fringes. In this case, after rotating the interferometer, the phase shift should change.

The theory of the Michelson experiment states that when the interferometer is rotated, both the mirrors and the directions of the light rays will change

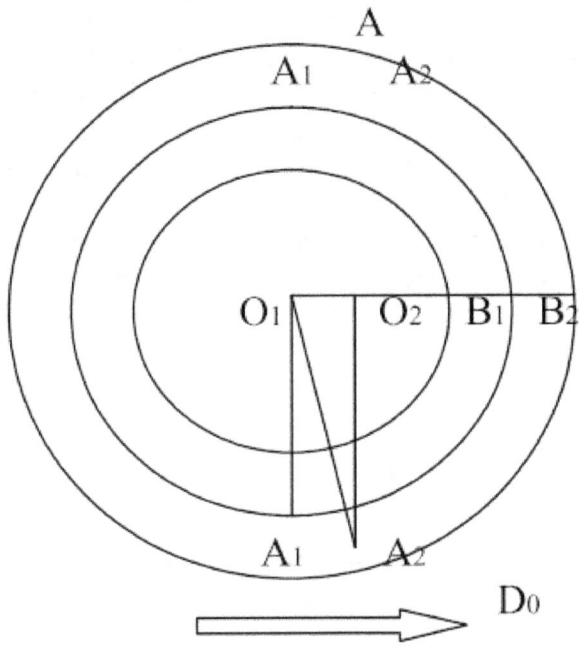

Picture 5.

places, thereby the shift of the fringes should double.

But does the interferometer have the ability to determine which beam and from which mirror came first or last? Has anything changed in the relative arrangement of the light wave and the mirrors? On fig. Figure 5 clearly shows that mirror **A1** will appear in place of mirror **B1**, and at the bottom of the figure, mirror **B'1** will appear in place of **A'1**. The mirrors have really changed places. But what happens to the waves of the light flux and the phases of their arrival? In both directions, the propagation of a light wave occurs according to the same law as before. It would be possible not to rotate the interferometer at all, but simply move the mirror from **A1** down to position **A'1**. For the interferometer, it does not matter which side the light flux comes from, the absolute value of the phase shift is important for it, and it, both in the first position and after rotation, will always be the same.

A very interesting case is the rotation of the interferometer by 45°. Then the paths traversed by the light wave along both of its arms will be the same. The phase shift between the two light fronts will be zero and it is possible to observe fringe interference. However, the length of the path will be less than in the first case along the line **O1-A2-O2**. Let us consider the position of only one arm of the interferometer from its upper position, which we will take as zero L90.

In this position, the largest phase shift is observed, at which the initial fringe interference pattern is observed.

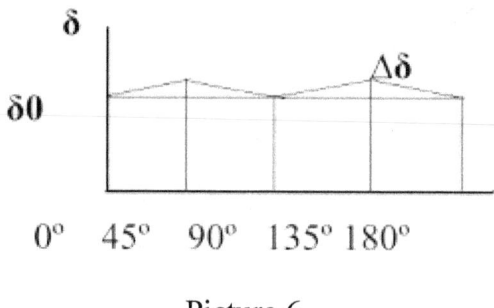

Picture 6.

When rotated by 90°, the same interference pattern and the same maximum phase shift of the two light signals are observed. There is no relative shift of the bands between these two positions. Let us take this phase shift as the zero state δ0. When the interferometer arm is moved by 45°, there is a relative decrease in the path length of the light beam L45, and hence the phase shift of the signal to a certain level:

$$L90 - L45 = \Delta\delta$$

This will make it possible to observe the shift of the interference fringes, but it will be so insignificant that it goes beyond the accuracy and all the errors of the experimental setup.

Consequently, there could not be any shift of the interference fringes in such a formulation of the experiment, as in subsequent ones. This is convincingly proved by the Michelson-Morley experiment. But the conclusions that were drawn from his results had enormous consequences. First of all, the theory of the ether was buried as a medium for the propagation of an electromagnetic field. Then, the universality of the speed of light was postulated, regardless of the motion of the system. And already on the basis of this, SRT and GRT were created. The worldview of the earthly humanity fell into the trap of space time curvature. There is only one way out of this, to recognize the failure of the results of the Michelson-Morley experiment.

ANNOTATION

«Physical foundations of force interaction».

An analysis of the historical concept of force shows that its physical properties, such as the absoluteness of magnitude and invariance in space, as the body moves, were preserved until the creation of Einstein's theory of relativity. At the same time, introducing the concept of a force field and using it in calculations, no one has ever tried to explain what its physical essence is. This was the main reason for limiting the application of Newton's equations only for cases of motion of bodies with low velocities. However, a new approach to the process of force interaction between two bodies makes it possible to eliminate the identified inconsistencies.

Considering rectilinear and circular motion in the field of the central force, their relationship is established and it is noted that any motion is a consequence of force balancing.

It is concluded that it is possible to accept the absolute value of the body's motion dynamics and reject the concept of inertial reference systems.

A number of experimental data are presented to illustrate the basic provisions of a new view on the process of force interaction of bodies.

«PHYSICAL BASIS
FOR FORCE INTERACTION»

INTRODUCTION

Simplicity and amazing accuracy in describing the trajectories of the mechanical motion of various objects, up to space ones, speak of a solid foundation for Newton's theory. Even when solving the problem in other ways, such as, for example, through the Lagrange equations, one has to set the type of functions in such a way that they still turn into a set of Newton's equations. In this case, there is absolutely no criterion for choosing functions, except for one – to obtain a solution and calculated data corresponding to the experimental data. However, the multitude of ways in which problems can be formulated and solved, their increasing mathematical complexity, do not reveal the physical essence of phenomena. Until now, the main question remains unanswered, what is power? The situation is analogous with the concept of the field and its physical nature.

This means that any of the theories describes only the external manifestation of force interaction through the movement of material bodies, based on some initial concepts and prerequisites. In this case, Newton's theory has all the advantages over the rest. Originating on the basis of experimental data for low velocities of bodies, it ceased to satisfy the new experimental data when these velocities increased significantly and approached light. Consequently, the question arose about the limits of applicability of Newton's theory and what it is connected with. Isn't this restriction purely conditional, which arose as a result of the formation of basic concepts? Was there a misinterpretation of the results of the experiment, or were there incorrect and redundant premises?

And, if so, what should the mathematical formulation of the law of force interaction look like?

These are the main questions that an attempt is made to answer in this article. In this case, the concept of the dynamics of the relative motion of the body and the RSCS coordinate systems will be used. (12).

ORIGIN AND DEVELOPMENT
THE CONCEPT OF FORCE

The well-known formula for Newton's second law of dynamics is:

$$F = ma$$

later, in the special theory of relativity (SRT), it passes into a more complex relationship:

$$F = m_0/\sqrt{(1- \beta^2)} \times a \qquad \beta = V/C \qquad 1.2$$

This raises serious and contradictory difficulties in its interpretation. Previously, the increase in force, with increasing speed, was explained by an increase in the rest mass of the body m0 and inevitably ran into the question: «Where does this increase in mass come from?» It was impossible to answer it, even using the idea of the relativity of this increment. At present, explanations are of a more cautious nature, and their physical content is significantly different. Thus, when determining the mass tensor, SRT says

«...the answer to the question posed about the experimental discovery of the dependence **mß** is to determine the individual components of the **maß** tensor». (1.2).

And it is concluded that in experiments it is easier to see only confirmation of the relativistic equations of motion. In fairness, it is noted that in particular cases, when the equations of motion are similar to Newton's, then, as if, the mass of the body changes due to the presence of speed.

Other explanations are associated with some external manifestations.

«It seems to us that this formula for changing the mass of a body with increasing speed reflects the interaction of this body with its environment, and not the so-called increase in the amount of matter and not the transformation of motion into matter or field». (3).

As you can see, none of the interpretations reveals the essence of the physical phenomenon. But at the same time, new questions arise, which are also not easy to answer. Therefore, let us turn, first of all, to historical facts.

The concept of force originated in ancient times and, above all, on the basis of atomistic theories around the 5th century BC. Neither Democritus, neither Epicurus ascribed to atoms certain forces of interaction at a distance. But at the same time, they pointed out that «like strives for like» (4). Lucretius explained this by the exchange of flows of invisible «light bodies», and the selectivity of interaction, as in the case of iron and a magnet, by the correspondence of the pores and bodies of these particles. Since there was no concept of force in the theories, almost all authors agreed that atoms, initially, had to fall in a vacuum under the influence of their gravity. Epicurus even pointed out that the velocities of falling bodies should be the same in the direction «to the bottom of the sheer». It is directly indicated that the movement is caused by some cause, and its magnitude (gravity) is constant and does not depend on the movement itself, since the atom itself is eternal and unchanging.

Criticizing the atomists, Aristotle comes to the conclusion that *«...everything that moves must necessarily be set in motion by something».* (5). He calls it dunamis power and refers to the action of one body on another. At the same time, a constant character in space and time is also attributed to the absolute magnitude of the force: «...*if A is moving, B is movable, G is the length by which it is advanced, D is the time during which it moved, then, in equal time, a force equal to A will move half of B by double the length of D, and by the whole of D in half the time of D, such will be the proportion*».

Based on the same premises, Aristotle derives his law of falling bodies:

«*If a certain weight travels a certain time, then a larger body will cover the same distance in a shorter time, and the ratio of the weights to each other will correspond to each other's times, i.e. half a weight travels a distance in X, then a whole weight travels that distance in 1/2X*». (4).

He attributed this increase in speed to «*a gradual increase in weight as the body approaches its own place*». Aristotle did not yet identify the concept of force with weight, but pointed out that the rate of fall depends on the medium.

We find a more meaningful concept of force in the ancient Chinese treatise on physics «MO CI»:

«*Force (li) is what makes objects that have a shape move*».

And further:

«*Heavy is strength. The fall of one object or the rise of something else is a movement caused by gravity*». (4).

But here, too, force is associated with movement, it is she who is the cause of the latter. There is a clear chain – «an unchanging force – the movement of the body», in all ancient physics.

Of particular interest is the first link – «unchanging strength». This property of hers was preserved until the creation of the SRT.

«Physics» of Aristotle has received wide recognition and distribution. However, it was subjected to serious criticism from other philosophers, whose works significantly deepened the concept of force, laid the foundation for its modern understanding.

Thus, in the commentator of Aristotle, Alexander of Aphrodisias, we find the following definition of force:

«...a moving body, at the moment of the beginning of the movement, receives a certain driving force, which supports the movement (impeto), when this force dries up, then the movement stops». (6).

This was connected, first of all, with attempts to explain how an abandoned body or an arrow shot from a bow moves. After all, there is no longer constant contact of bodies in such cases. A more interesting explanation of such a movement is given by Philopon in his «Comments on Physics»:

«Or rather, it should be that the thrower imparts to the thrown object some incorporeal power – kinetic, but the air that he pushes does not add anything or very little to this movement». (4).

Force, as a concept and some physical reality, begins to acquire the contours of some independence, it is no longer associated with the contact of bodies. Philopon also criticized Aristotle's law of falling bodies:

«If we make two bodies, sharply different in weight, fall simultaneously from the same height, then we will find that the ratio of the times of their movement is not equal to the ratio of their weights, but that the difference in time is very extremely small». (4).

On the basis of such an understanding of force that it is the cause of motion, the most interesting question was to what speed the motion will increase if a constant force acts on the body for an infinitely long time. Indeed, in a vacuum, where there is no resistance of the medium, the speed would have to increase to infinity. This clearly contradicted conventional observations. Then the body got would be the property to instantly appear and disappear, i.e. omnipresence.

Aristotle rejects emptiness and introduces the concept of environmental resistance.

The ideas of Aristotle dominated physics until the 17th century. (4). At the same time, as the theory of body motion develops, concepts of instantaneous speed and acceleration are formed, and the path is determined for uniformly accelerated motion. J. Buridan sharply criticizes the provisions of Aristotle as inconsistent with experimental facts. The main problem of mechanics, the relativity of motion, is being investigated. Thus, the ground is being prepared for Galilei's new worldview. A fundamental clarification was introduced into the concept of force, thanks to a discussion on the problem of the kinematic and dynamic description of movements. The meaning of this was as follows. On the one hand, it was considered necessary to describe the movement taking into account the causes that cause it, i.e. forces. On the other hand, it was proposed to consider the movement as a pure movement. For the first time this issue was studied in detail by T. Bradwardine. (4). He divided «the proportion of speeds in movements, in connection with the forces of movers and movable objects, in relation to the magnitude of movable objects and the distances they cover». R. Swainshead also adhered to an analogous opinion, separating the study of movement in relation to its cause and in relation to its result.

But the question – to what extent the speed increases under the action of a constant force that is unlimited in time, remained unanswered. Criticizing Philopon, Bradwardine puts forward the idea that the ratio of the speeds of movement follows the ratio of the driving force to the force of the object in such a way that

«...if the speeds V grow in arithmetic progression, then the ratio F / R in geometric (F – force, R – resistance)". (4). Mathematically, this could be written as»:

$$F2/R2 = (F1/R1)^n, \quad \text{where} \quad n = V1/V2$$

He specifically notes:

«No matter how great the force, if the resistance grows to equality with it, and the force remains unchanged, then under the

influence of this, the body will move at a speed approaching infi-nite slowness».

However, he does not reveal what the nature of this resistance is, what causes it. The action of force and resistance in its inter-pretation is not yet interdependent, as in SRT.

He says only that if the resistance increases to the magnitude of the force, then there is an approach to infinite slowness. Critics immediately raised questions, but if it does not increase to this magnitude of force or increases according to some other depen-dence, what then? It is noteworthy that already at this stage, when the observed speeds are small, the problem of the limiting speed to which the body can be accelerated under the action of a force of constant magnitude arose with all its acuteness.

Thus, the concept of force was based on two fundamental principles. First, force is the cause of motion and how its mani-festation, and also, in a sense, its quantitative assessment is speed. The second – the magnitude of force, as a dynamic con-cept, was determined in the same way as in statics, by a constant number for each specific case. Research by other physicists during this period did not bring any great ideas into these princip-les. (4, 6, 7, 8).

In the 16th century, by the works of Albert of Saxony and Nikola Oresme, a classification of all movements was made, its graphical representation was introduced, and a modern definition of speed was given. As a result, new ideas about power begin to emerge. And, the first of them, we find in Benedetti in his main work «Various mathematical and physical reasoning» of 1585. Here we already have the foundations of the principle of inertia, which he used to explain the motion of bodies. By itself, some-one's interpretation of the concept of force is not new:

«...the constant increase in the speed of falling bodies is due to the accumulation of the action produced by the same cause of motion, and not to a gradual increase in weight, as Aristotle said». (8).

Approximately the same is found in Buridan, approximately two centuries before him:

«...the engine, setting the moving body in motion, introduces into it a certain pressure or some kind of driving force inherent in the moving body». (4).

Already Benetti associated the action of force with a change in speed, i.e. acceleration, and this is his great merit. Arguing further about the movement of the various spheres relative to the center of the Earth, he uses the original proof, dividing the spheres into equal parts and concludes:

«...two bodies of the same shape and the same kind or not equal to each other, in the same medium pass equal distances in equal time». (8).

These views were adopted by Galileo. His era was already characterized by a transition from philosophical discussions to experimental research and theoretical generalizations. Together with the ideas of Hilbert and Kepler, in physics, the idea of force as a single law of nature that governs the solar system begins to assert itself. Not only was an anology of the force action between gravity and magnetism revealed, but thoughts were also expressed about their action at a distance, without anything material in the intermediate space. (4).

The main task of Galileo was the study of natural motion – the fall of bodies. He noted:

«...the natural motion of a falling heavy body is continuously accelerating. However, in what respect this acceleration occurs has not yet been indicated». (4).

And he finds this ratio thanks to experiments with an inclined plane. Installing it at different angles, and measuring time by the amount of water flowing out of a hole in the bottom of a suspended bucket, Galileo found that «the ratio of the paths traveled is equal to the ratio of the squares of their travel time». It should be noted that the concept of acceleration was not yet known to Galileo. Therefore, he gives the following definition of uniformly accelerated motion:

«I call uniformly accelerating motion such a motion in which, starting from a stationary state, the speed, at the same time intervals, increases equally». (9).

Of fundamental importance is the fact that he associates the action of a force of constant magnitude with a change in speed, and not with the speed itself.

A remarkable moment in the history of the development of the doctrine of the movement of bodies and the disclosure of the mechanism of action of force is the principle of inertia. He eliminated the need for a continuous force on the body to maintain its movement in space. The body has acquired the ability to move independently without any external mover. At the same time, it was always emphasized that its movement is uniform and rectilinear. Galileo came to an understanding of inertia through reflections on the transition of the pendulum from falling to its rise:

«...a ball suspended on a thread, after passing the lowest point, rises almost as much as it fell».

He considered it in relation to a smooth surface «which does not rise or fall»:

«When a body moves along a horizontal plane without encountering any resistance to movement, then ... its movement is uniform and would continue constantly if the plane extended in space without end». (4).

The movement of the body does not need a force constantly acting on it – this is the main conclusion of Galileo.

Carrying out experiments with an inclined plane and a pendulum, Galiley came to another, no less important conclusion:

«...if the resistance of the medium were completely eliminated, all bodies would fall at the same speed». (8).

This was the most important result of his research. Here is how Galileo himself explains the course of his reasoning regarding the fall of a stone:

«If it is clear from the very nature of things that the speed cannot remain the same and that the movement cannot be uniform, then we must look for constancy, or, if you like, uniformity and simplicity, not in speed, but in the increment of speed». (9).

Thus, Galileo discovered and determined the force of gravity:

«Deduce from this conclusion that in a free and natural fall, a small stone does not press on a large one and, therefore, does not increase its weight, as happens when it is at rest». (7).

Having understood this, Galileo concludes that the acceleration of gravity is independent of weight. The only reason is that it is necessary to take into account the resistance of the environment. Based on this understanding, a general idea of dynamic force arose, and therefore a new doctrine of motion. Galileo does not give a formula for force or its definition, but his worldview formed the basis of Newton's research.

Summing up some of the idea of force, in its development before Newton, we can single out the following fundamental points:

a) uniform and rectilinear motion of the body does not require the application of any force;

b) a force of constant magnitude causes equal increments of speed;

c) the possibility of a force acting on bodies without an intermediate medium;

d) universality of the gravitational force in relation to all bodies of any size and weight, which is expressed in their equal magnitude of change in speed.

As can be seen from this, the concept of force has undergone significant changes since the time of Aristotle – from the «manpower» of slaves to objective physical reality. Without disclosing the physical essence of the force, the main provisions in the nature of its effect on the body were changed, which is expressed, as the final result, by a change in the speed of the body, which was recorded during the experiments. However, the magnitude of the force itself again remained constant. Did not find its reflection in

110

the studies of Galileo and the question – to what extent will the speed of the body change with the constant action of a force of the same magnitude on it? After all, according to the definition of force, the speed of the body would have to increase to infinity. How could this apparent inconsistency be related to actual observations?

Perhaps the first who tried to explain this discrepancy was R. Descartes. In one of his letters, he wrote:

«It can be stated with certainty that a stone is not equally disposed to accept a new movement or to increase speed when it moves very quickly or when it moves very slowly». (7).

However, his statement was just a brilliant guess, which went unnoticed. Rejecting the idea of the immutability of atoms, he expressed a number of thoughts about the interaction of particles, including the variable value of inertia. Reflecting on the effect of force on a particle, he noted that the result depends on its state, i.e. not only on the amount of matter, but also on its speed. Descartes did not seek to develop some kind of mathematical theory on the basis of his ideas and derive some new consequences from this. This, unfortunately, also contributed to the fact that his views were simply condemned and forgotten. In the concept of force, Descartes did not introduce anything new and fundamental. Weight, like force in general, he considered as a reaction of a geometric type. Having filled the space with whirlwinds, like the movement of some subtle matter, it could be argued that weight and force are a property of space. But this hypothetical construction only complicates the understanding of the world and excludes the possibility of deriving mathematical laws. Therefore, Cartesianism quickly gave way to Newtonian dynamics. (8).

Having introduced the definitions of mass and momentum, Newton, further, in his famous «Principles» proceeds to the definition of inertia and force. Let's consider these concepts in more detail.

«Definition III. The innate force of matter is the ability of resistance inherent in it, with the help of which any single body, since it is left to itself, maintains its state of rest or uniform rectilinear motion» (4).

As follows from the explanations of the commentators, by innate power here is meant the passive power of perception. Dale Newton adds:

«This force is always proportional to the mass, and if it differs from the inertia of the mass, then only by looking at it».

Thus, inertia is determined only by the mass of the body, and therefore is a constant value.

«Definition VI. An applied force is an action performed on a body in order to change its state of rest or uniform rectilinear motion». (4).

He emphasized:

«Power manifests itself only in action and does not remain in the body after the action has ceased. The body then continues to maintain its new state due to inertia alone. The origin of the applied force can be different: from impact, from pressure, from centripetal force». (8).

Interestingly, he attributed the force of gravity to the centripetal force. At the same time, one definition of force in this case is not enough, since the force acting from the same source at a different distance from it has a different value. Therefore, Newton characterizes the very source of the centripetal force as an «absolute value». And the force that acts on the body at a given point in space is the «accelerator value», which corresponds to the applied force, but calculated per unit mass or a driving force equal to the force for a given distance from the center.

«I do not understand here how these attractions can be carried out ... I use this word here to designate in general a certain

force due to which bodies tend to each other, whatever the reason». (4)

Having assigned a certain absolute value to the source of force, he found a method for quantifying the force of any physical nature through the movements of bodies. And he came to this, first of all, through the law of the fall of Galileo. The most important thing is that identical physical prerequisites are laid down here, namely:

a) for each specific case, the magnitude of the force has its absolute value;
b) this force causes the same increment of speed throughout the entire motion of the body;
c) unlimited value of final velocity.

But, as you can see, there are differences. Newton is talking not just about force, but about the applied force for each specific moment of consideration of its effect on the body. Hence Newton's second law states:

«The change in motion is proportional to the applied driving force and occurs in the direction of the straight line along which this force acts». (4).

It is characteristic that in his «Principles» and even rough drafts, Newton, with all the thoroughness of his formulations, did not give the mathematical formula of the law. Moreover, it is believed that he deliberately gave his laws of motion such formulations that should not have been translated into the language of formulas at all. The writing of the formula itself was done thanks to the commentators, who in the end, after long disputes and corrections, received a mathematical dependence. We are using it to this day.

In 1881, J. Thomson established the variable nature of the electromagnetic mass of an electron, and, later, N.A. Umov predicts that at speeds close to the speed of light, the mass should increase. Physics stood on the threshold of creating the theory of relativity. What results and views physicists came to in this case was already mentioned at the beginning of the article. Here we emphasize only the following. At this stage in the development of physics, not only guesses, but also experiment proved that any force can accelerate a body only to a certain level of speed. Moreover, the limit to which it would be possible to accelerate the body in general has become limited. In the theory of relative (SRT) it was determined thanks to experimental data and the introduction of special postulates. At the same time, neither the formula for determining the force, nor the nature of the very interaction of the force on the body, were revised in their physical basis. It can be said that, in this sense, Newton's formula has been modernized for the case of high speeds.

Although supporters of SRT claim that Newton's law is a particular case of a more general law. Apparently, this is still not true.

Let us define the main fundamental points that relate to the modern understanding of force, from the point of view of its physical essence:

a) the accelerating force is determined by the absolute value;
b) variable inertia of the body, with increasing speed;
c) variable acceleration, as a consequence of the variable inertia and the absolute magnitude of the force;
d) a certain value of the boundary speed, equal to the speed of light.

These are the foundations that are characteristic of force and its interaction with the body. But, as the analysis convinces, they are based on the law of falling bodies of Galileo. Well, what if his conclusions were not entirely objective, due to imperfection and limited experience? Let's make sure this is true. There is no doubt that his experiments with the fall of heavy bodies from the Leaning Tower of Pisa could not give a satisfactory result in terms of measurement error. (4). Nevertheless, by conducting these public experiments, he was able to refute Aristotel and his law of falling bodies. And, most importantly, he came to the conclusion that the speeds of falling bodies are the same and independent of weight.

114

«A grain of sand must fall with the same speed as a millstone», said Salviati. Oddly enough, but this led him, in the future, also to another statement – the same increment of speed throughout the fall of the body, which is fundamentally different from the first. It is one thing, the same speeds of movement of bodies in relation to each other, which was fixed approximately at the moment of their impact on the surface of the earth, and quite another is the constancy of the increment of their speeds during movement. After all, the increment of speed can be different, but the relative speeds themselves remain the same in magnitude. We can only talk about the same speed increment for different bodies, but not about the constancy of the speed increment for the same body during its accelerated motion.

And only after that, Galileo performed experiments with an inclined plane and determined the quantitative dependence. The length of the trough, along which the bronze ball rolled, was approximately 7.2 m, the accuracy of time measurement was up to 1/800 of a minute. (4.7). By changing the inclination of the gutter, he made sure that «the ratio of the paths traveled is equal to the ratio of the squares of the time they traveled».

Combining this with the statement about a constant increase in speed, Galileo comes to the well-known formula for the fall of bodies. But, as can be seen from the description of the technique of the experiments themselves, they were very limited and did not have sufficient measurement accuracy. First, the statements refer to very insignificant heights, and therefore to the path that the body travels. Secondly, the speeds that the ball acquired when rolling down the chute were also very small, and it was impossible to detect a possible deviation from the constant value of the acceleration of gravity.

«In any case, it becomes quite clear that it is not experience that makes us believe in the unshakable force of the laws of fall, we cannot even consider them otherwise, as only the first approximation to the establishment of the first component during movements on the surface of the Earth. ...We are disposed to them by the contrast between the simplicity of laws and the complexity of phenomena, to the explanation of which he gives the key». (9).

On the other hand, the theory of relativity seems to indicate the possibility of a variable acceleration when a body moves under the action of a force of constant magnitude, but it did not fully reveal the whole essence of physical phenomena. Various interpretations and an outwardly witty explanation of the effects only indicate that, apparently, some inaccuracies were made in the very understanding of force and its interactions that eluded researchers of the times of Galileo and Newton. Having accepted their axiomatics, modern scientists have been forced, in order to harmonize theory and experiment, to introduce additional prerequisites and build more complex mathematical models. This led to the fact that very strange uncertainties appeared in calculations close to the speed of light, as well as in non-linear coordinate transformations. Is it justified? Apparently not.

BASIC RELATIONSHIP FOR FORCE INTERACTION

From the analysis of historical facts, it is necessary to highlight two points that are mutually exclusive. On the one hand, this is the absolute value of the very concept of force, and, consequently, its static and invariable in space, as a body moves. And on the other hand, the limited impact of this force, which is expressed, ultimately, in the maximum speed of the body.

Therefore, the first question that needs to be answered is whether it is possible at all to combine such concepts as the absolute and constant value of force in space and the acceleration of a body with a finite speed limit.

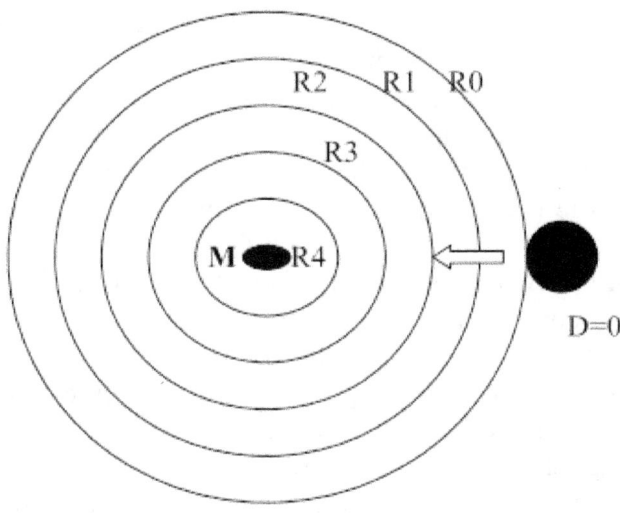

Fig.1

Let us have in space the center of force of the cape **M**, Fig.1. The gravitational field, through which the effect of the center of force on a body of mass **m (m<<M)**, will be represented as elementary spheres of radius **R**. We will assume that at a distance **R0** the gravitational field of the center of force will be equal to zero, and the center itself is motionless. What happens when the body moves through the elementary spheres of the gravitational field, if its initial dynamics is equal to zero **D=0**. First of all, under the influence of the center of force, its dynamics will increase uniformly by the value of **ΔD**. But, most importantly, at the same time, the magnitude of the force acting on the body, i.e., continuously increases. gravitational field. Since the change in **ΔF** is inversely proportional to the square of the radius, we can write:

$$\Delta F = kM/R^2 n \qquad 2.1$$

where **k** is the proportionality factor.

Newton, conducting his experiments with a cart, used a load and naturally identified it with a constant driving force. But the load is in the gravitational field of the Earth, and is a kind of «transmission mechanism» between the cart and the center of force. The small length of the path when the trolley moves along the surface of the Earth and its speed just leads to the conclusion that it is possible to cause changes in the speed of the trolley with the help of a constant force. However, Newton was just as well aware that the weight of a body changes with height. Apparently, this was the reason why he did not give the formula for the second law of dynamics. He could not find a mathematical equivalent to these two facts and was forced to confine himself to indicating the «driving quantity» of the applied force.

Let us now assume that the body, moving at an accelerated rate from **R0** to **R1**, acquired some dynamics **D1**. Further, beyond the sphere **R1** to **R3**, something incredible happens. Mentally imagine that in this area the gravitational field of the center of force does not change and remains constant in magnitude equal to the sphere **R1**. Will the body receive an increase in speed while moving in this area. Obviously not. It will continue to move by inertia with the same dynamics that it acquired when it left **R1**. Therefore, answering the first question, it is necessary to conclude that only a change in the magnitude of the force can lead to a change in the dynamics of the body's motion. It is fundamentally impossible, due to some static, absolute value in space, to cause changes in another value. This is tantamount to as if the entire mass of the center of force was evenly distributed in the entire surrounding space, which means that there would be no reasons for changing its dynamics.

Recognition that it is not the absolute magnitude of the force that causes changes in the dynamics of the body, but its increments, as it moves, allows us to take a new approach to the question of the final magnitude of its impact, i.e. about the dynamics that a body can achieve under the influence of a given center of force. All accumulated experience shows that a body of finite mass also has a finite gravitational field.

In this case, a direct relationship is also revealed, the greater the mass of the body, the stronger the gravitational field surrounding it.

Thus, it can be argued that the sum of all values ΔFn is also a finite value:

$$\sum_{n=1}^{n} \Delta Fn = Const \qquad\qquad 2.2$$

Consequently, the maximum dynamics that the body reaches, moving throughout the action of the center of force from **R0** to **R5**, is also finite.

Expression 2.2 was defined for a given center of force with mass **M**, Fig.1.

It is quite natural that for another center, with a different value of mass, there will be a different value of the sum **ΔFn**, and hence a different value of the final dynamics. Thus, we come to the second fundamental conclusion – for each center of force, there is its own maximum dynamics **Dm**, which a body can achieve, moving in its gravitational field with an initial dynamics equal to zero. But, how to correlate in space, precisely as the body moves, such concepts as the absolute, finite value of the center of force itself and its dynamics **Dm**, with other concepts, such as an increase in force and a change in the dynamics **ΔD** of the body?

Let's consider another thought experiment. Let there be a center of force **M**, around which there are only two areas of the gravitational field: **R0-R1** and **R3-R4**, Fig.1. A body of mass **M** will begin accelerated motion in the first region and, having passed the outer boundary **R1**, will reach some dynamics **D1**. Further, moving uniformly, in the area where the gravitational field is absent, it will reach the second area on the sphere **R3** and will inevitably start accelerating again. But will this change in the dynamics of **ΔD2**, in absolute value, correspond to the value that was in the outer region? Indeed, in this case, the value of the increment of force **ΔF2** can remain constant, but it is, as it were, at a different level of its power, since there is a gap between the first and second levels. The body will be affected by different magnitude force impulses. The concept of Newton's «driving force» equalizes them in all respects. Apparently, in these two cases, the impact of the force exerted on the body by the central force will not be identical. The measure of this can be the magnitude of the difference in the increment of the dynamics: **ΔD2 – ΔD1**.

This is also caused by the fact that in the first case, the initial value of the dynamics **D0** and the process of its change, for a given sphere of the central force **ΔF1**, will occur according to one law, and in the second case **D2** and for the same difference **ΔF2**, but of a different level, this law will inevitably look different, since the initial conditions themselves are already violated. In reality, in space, around the center of force, there is a continuous succession of elementary spheres of the gravitational field one after another. After passing one such sphere, the dynamics of the body, in general, will change from **Dn-1** to **Dn**, i.e. can be written:

$$\Delta Di = \Delta Dn - \Delta Dn\text{-}1$$

Since, each subsequent elementary sphere can be considered as a center of force with a value:

$$Fn\text{-}1 = M/R^2 n\text{-}1$$

Then, accordingly, for any of these centers, there will be its own value of the increment of body dynamics:

$$\Delta Dn = Dn - Dn\text{-}1$$

where Dn is the dynamics of the body, which it will acquire by moving in the field of the given center of the elementary sphere **Fn-1**;

Dn-1 – initial dynamics of the body at the input of this elementary sphere.

Moving on, further, directly to the center of force **M** and the movement of the body throughout the action of the gravitational field, we can write the ratio of dynamics for any of its positions, as:

$$\Delta D'n = DM - DM\text{-}n$$

where **DM** is the maximum or boundary value of the dynamics that a body can acquire when moving in the field of the center M;

DM-n is the value of the dynamics that the body has acquired on any of the elementary spheres.

The physical meaning of this lies in the fact that this difference determines the reserve» that still remains at the center of force, to maintain changes in the dynamics of the body during its accelerated movement.

However, for different centers of power, the ratio will have different values. In order for it to take on a universal character and this could be taken into account in the general formula for any centers of forces, it is necessary to know the unit ratio of dynamics, i.e. normalize by the value of **DM**. Then expression 2.1 will be written as:

$$\Delta F = \frac{DM - \Delta DT}{DM} \times m \times \Delta Dn = (1 - ß) \, m \, \Delta Dn \qquad 2.3$$

It was already mentioned earlier that the value of Δ Fn has a different value each time, i.e. another level of reference. What is he equal to? The body, successively passing one elementary sphere after another, will change its dynamics from zero to some value **DT**. In this case, the change in the magnitude of the force occurs to the level:

$$Fn = \sum_{i=1}^{n} \Delta Fi$$

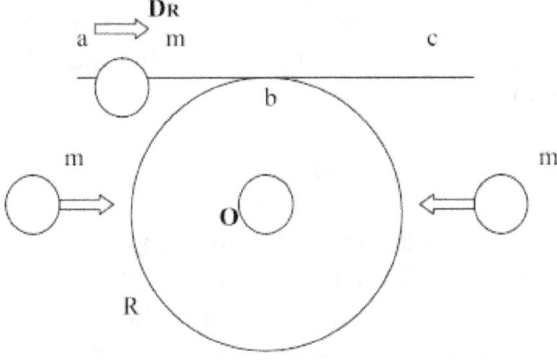

Fig. 2.

Therefore, for any of the spheres of the gravitational field, the following equality can be written:

$$Fr = m\ Dr \qquad 2.4$$

The importance of this ratio lies also in the fact that it is with such dynamics that the body will move along the circular orbit of the given sphere. Indeed, let there be only one sphere of the gravitational field with a value around the center of force **O**, Fig. 3:

$$Fr = kM/R^2$$

If a body of mass m moves uniformly and rectilinearly at a distance **R** from the center **O**, with dynamics **Dr**, then it will move to a circular orbit. It will remain in this orbit until its dynamics exceeds **Dr**. If we assume that the sphere of the gravitational field suddenly disappeared, then the body will again continue to move uniformly and rectilinearly with the same dynamics. However, in the case of a circular motion, it is argued that Newton's third law is fulfilled in this case, i.e. there is a balancing of two forces – centrifugal and centripetal. Consequently, the value of **mDr** is equivalent to the force effect on the sphere **R** and does not depend on the direction of the body motion. But the same magnitude of force, in this case, must retain its inner content and significance also for rectilinear motion. If the same central force acts on bodies with mass **m1 << m2**, then as a result their dynamics will change and will be determined, in general, by the values **D1>>D2**. However, in relation to the very level of force impact, their movements will be completely different. This means that their values are equal.

$$m1\ D1 = m2\ D2,$$

not by chance and identically, in the sense of the equivalent of force. This is the fulfillment of Newton's third law in relation to rectilinear uniform motion. It confirms the fact that there was a balancing of two forces at some level, and the body we observe carries a «memory» of the magnitude of this interaction. In this case, as can be seen, the magnitude of the body's motion dynamics will unambiguously characterize both rectilinear motion and motion along a circular orbit.

Thus, to find the total value of the central force, in the case of an accelerated motion of the body, it is necessary to write down:

$$\mathbf{Fr} = \Delta\mathbf{Fi} + \mathbf{Fr}$$

Since the expressions for Δ **Fi** and **Fr** are valid for any sphere of the gravitational field of the central force, then for a body of mass m we will have:

$$\mathbf{F} = (1 - \text{ß}) \ \mathbf{m} \ \Delta\mathbf{Dn} + \mathbf{m} \ \mathbf{Dr} \qquad 2.5$$

This expression allows us to explain in a new way what determines the limitation of the maximum dynamics of the body during its accelerated movement towards the center of force. As **Dr** increases, there is a decrease in the effectiveness of the influence of the center of force on changing the dynamics of its movement. The value of the center of force itself is absolute and the value of **Dr** = **Dn** (the limiting value of the velocity for the body) does not depend on changes in any, even relative, properties of the body, such as its mass.

Consider the general case when the initial dynamics of the body D0 is not equal to zero, Fig. 4

Thus, the body already had its own «history» of movement in the field of some other central force. If the body had an initial dynamics **D0** equal to zero, then moving in the field of the central force from **R0** to **R1**, it would acquire the same value of the dynamics **D'0**. Since the initial dynamics of the body **D0** is not equal to zero, it is possible to assume that the body moves in the field of some center of force from the initial point **R0** to its final value Rc, while acquiring a general dynamics equal to:

$$\mathbf{D'c} = \mathbf{Dc} + \mathbf{D'0}$$

In this case, the body reaches its maximum dynamics Dt in the force field of a given force. However, the body had an initial speed, and consequently, the effectiveness of the impact of the center of force on its acceleration will be less and will be:

$$D'c = \frac{Dc}{Dc + D0} \times Dc$$

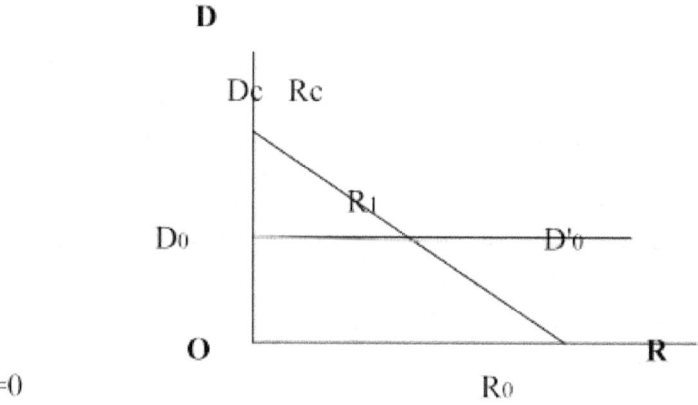

Picture 3.

Since the increment of the body dynamics will occur relative to the reference level **D0**, i.e. from the point of the sphere **R1**, then substituting the values **D'c** and **D't** into expression 2.5, we get:

$$F = \frac{\dfrac{Dc}{Dc + D0} \times Dc - (D_T - D0)}{Dc} \times m\,\Delta D + mD_T =$$

$$= \frac{D'c - D'_T}{Dc} \times m\,\Delta D + mD_T \qquad 2.6$$

124

It follows that any body with an initial speed will move in the field of the central force up to the maximum value of the dynamics equal to:

$$D_T = D0 + \frac{D^2c}{Dc + D0} \qquad 2.7$$

It can be seen that for **D0 = 0** formula 2.6 goes over into 2.5. If, for example, a body moves with the initial dynamics **D0 = Dc**, then it will receive the maximum increment of dynamics **0.5 Dc**, and its final dynamics will be **Dt = 1.5 D0**. The higher the value of **D0**, the less change in dynamics will cause a given center of force, the less influence it will have on the movement of the body.

It is well known that if the shape of the center of force is not spherical, for example, an ellipsoid of revolution, then this will also lead to a «deformation» of its gravitational field, in comparison with a spherical one.

This will have an effect on the difference in the acceleration of the body, as if it were moving along the major axis of the ellipsoid or along the minor one. However, the deformation of the gravitational field can occur not only due to the difference in the shape of the body of the central force or the distribution of mass density, but also due to the movement of the center of force in space, with some dynamics **Dc**. By itself, the gravitational field, as one of the forms of the existence of matter, inevitably has its own measure of inertia and it is possible to attribute its own value of the dynamics **Dgr** to it. Consequently, the magnitude of the deformation of the gravitational field during the movement of the center of force, and hence the degree of influence on the change in the dynamics of the body, can be determined through the ratio **Dc/Dgr**.

Then finally we will have:

$$F = (1 \pm \frac{D_{\text{и}}}{D_{\text{гр}}})(\frac{D'c - D'_T}{Dc}) \times m\,\Delta D + mD_T \qquad 2.7$$

125

This expression is valid for any center of force and any initial dynamics of the body. Knowing the magnitude of the dynamics **Dc, Dᴎ, Dgr**, it is possible to estimate the degree of influence exerted by the central force throughout the movement of the body.

ROUNDABOUT CIRCULATION

As already noted, when a body moves in the gravitational field of a central force at **D0 = 0**, the value of its dynamics **Dt** at each individual moment for a certain distance R will correspond to the one that the body should have at circular motion of the same radius. In this regard, circular motion occupies a special place. Having an analogous characteristic as rectilinear motion, it allows one to estimate the magnitude of the dynamics in its absolute calculation in relation to the central force. This becomes possible due to the symmetry of the gravitational field of the central force and taking into account its deformation during the movement of the center. It also becomes possible because the force interaction depends only on the magnitude of the body dynamics **Dt**, relative to the center of force itself.

But this means that the already uniform translational movement of the center of force and the body itself with the same dynamics will affect the nature of the circular motion of the body, i.e. its dynamics in orbit around the center of force. In this case, deformation of the spheres of the gravitational field will be observed. This will lead to a violation of the equilibrium state of the force interaction as the body moves from point **A** to point **B**, Fig. 3. The value of the radius of the orbit will change to a smaller side, therefore, the value of the dynamics of the body will increase.

The constancy of the dynamics of the circular motion of the body, with a fixed center of force, indicates its absolute nature. Thus, both the value of the dynamics of the body **Dt**, and the value of the dynamics of the center of force **Dc** can be found and used as constant absolute values. Taking into account that during circular motion $\Delta D=0$, and the magnitude of the force interaction, at a given level of the sphere, is proportional to **Mm/R²**, we have for two different spheres the same center of force, based on 2.7:

$$kM/R^21 = Dr1 \quad kM/R^22 = Dr2$$

where:

$$Dr1 \ R^21 = Dr2 \ R^22 \qquad 3.1$$

Knowing the value of one of the dynamics for a given circular orbit, it is possible to determine any other orbit of any radius.

From 3.1 it can be seen that the smaller the dynamics of the circular motion of the body, the larger the radius of the sphere corresponds to it, and the smaller the force is needed to keep this body in a circular orbit. However, for practical needs and to simplify calculations, it is more convenient to use another ratio:

$$D1 \ R1 = D2 \ R2 \quad 3.2$$

where

$$D1 = \sqrt{Dr1} \qquad D2 = \sqrt{Dr2}$$

Similarly, as for 3.1, you can get the ratio for the transition from one orbit to another:

$$\Delta R/R = \Delta D/D \qquad 3.3$$

D is the dynamics of the body in a circular orbit of radius **R**;

Δ D is the increment of body dynamics corresponding to the change in radius **Δ R**.

It is quite obvious that if at least one body revolves around a given force in a circular orbit, then it is possible to determine the value **Dc** of the center of force. Thus, the calculation according to formula 3.3 for the Sun-Earth system gives the value of **Dc** for the Suna equal to 214.9, which corresponds to a speed of 437.9 km/sec. The same calculation for the Earth-Moon system gives the magnitude of the dynamics for the Earth equal to **Dz** = 61.54, while **V** = 8.025 km/sec. This is exactly the value of the dynamics that the body acquires, moving rapidly in the gravitational field of the Sun or the Earth, from infinity at **D0 = 0**. If the movement of the body around the center of force is uneven in its orbit and the deviation **Δ D** from the average dynamics is known movement **D**, then 3.3 makes it easy to determine how much the radius of the orbit of the body **R** will change, and, consequently, calculate its eccentricity. For example, it is known that the average speed of the Earth's orbit is n = 59' per day, and its deviation is ±2'. Then the calculation according to 3.3 respectively gives **R** = 2.5 10 km, and **e** = 0.0167. It is also possible to calculate any orbit of the planet and its motion dynamics **Dp**.

Taking the radius of the Earth and its motion dynamics as single ones, we will obtain data for all the planets of the solar system, which are given in Table No. 1.

Since, in expressions 3.2 and 3.3, the characteristics of the center of force, such as mass and dynamics, are not included, then these relationships must be fulfilled for any other central force. So, for some satellites of Jupiter, taking the dynamics of Amalthea equal to one, we obtain the corresponding values of **Dsp**, which are summarized in Table No. 2.

Obviously, the value of the speaker from tables No.1 cannot be directly compared with the data from the speaker from tables No.2. Formulas 3.2 and 3.3 define only a general pattern for circular orbits of a given central force, in relation to each other. Taking one of them as the origin and assigning to it the corresponding value of the dynamics, the dynamics of all other bodies are found. In order to then compare them with the dynamics of the movement of bodies around another center of force, it is necessary to bring them or normalize them with respect to the central force of interest. Then, similarly, as in case 3.1 for centers with masses **M1** and **M2** we will have:

$$\textbf{M2/M1 = R1 D1/R2 D2} \qquad \textbf{3.6}$$

So, if we have determined that the dynamics of the Moon in the circular orbit of the Earth is equal to unity, then the same dynamics in the orbit of the Sun will already be relative to the unit dynamics of the Earth: $Dl = 1.15 \cdot 10^{-3}$ which corresponds to $R = 12.8 \cdot 10^{10}$ km. All of the above considerations are valid for the cases

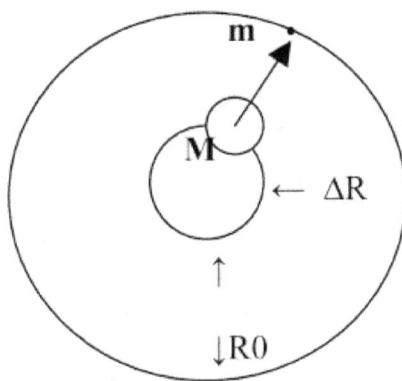

Picture 5.

when the mass of the center of force significantly exceeded the mass of the body and this allowed us to consider it motionless in the process of force interaction. In general, the center of force will also make a circular motion, fig. 5.

The distance between the center of force, mass M, and, the body of mass m is reduced by:

$$\Delta R = R0 - R$$

This will inevitably lead to the fact that the dynamics of the body will turn out to be somewhat greater, relative to how it would be if the center of force were stationary and always located at the same coordinate point. For the case of a fixed center of force M from 3.1 we have:

$$R0\ D0 = R\ D$$

For the moving center of force then it will be:

$$R0 \ D0 = R0 \ D - \Delta \qquad 3.7$$

Since the value of Δ is entirely determined by the movement of the center of force with mass M, then

$$\Delta = \Delta R. \ Dc$$

where Dc is the dynamics of the center of force.

The body m can also be considered as a center of force, in the field of which the body of the center of force M moves, then using the relation 3.6 we find:

$$\Delta \ RDc = m/M \ x \ RD$$

Substituting into formula 3.7 we get:

$$R = M/(m+M) \ x \ R0 \qquad 3.8$$

TABLE 1

PLANETS	$R \ x \ 10^3$ км	Dп	V км/сек
Mercury	57,9	2,57	47,9
Venus	108,2	1,38	35,0
Land	149,6	1,0	29,8
Mars	227,9	0,66	24,1
Jupiter	778,3	0,193	13,1
Saturn	1427,6	0,104	9,6
Uranium	2871,0	0,052	6,8
Neptune	4498,6	0,033	5,4
Pluto	5908, 9	0,025	4,7

TABLE 2

SATELLITE	R x 10³ км	Dп	V км/сек
Amalthea	181,3	1.0	26,46
Ganymede	1070	0,17	10,87
Lysitea	11710	0,0155	3,7
Ananke	20700	0,0087	2,44
Sinope	23700	0,0076	2,27

This ratio shows how many times the value of the radius will decrease during the circular motion of the body in the field of the central force with their mutual influence. It is this value for R that must be used in 3.6 for the general case. Consider an example for the Sun-Earth-Moon system. For the Moon's orbit we have:

$$Rl = Mz / (Mz + Ml) \times R'l$$

where **R'l** is the value of the radius of the Moon's orbit with the fixed center of the Earth.

For the Earth's orbit, respectively, we obtain:

$$Rz = Ms / (Ms + Ms + Ml) \times R's$$

Substituting the expressions for **Rl** and **Rz** in 3.6 we will finally have:

$$(Mc+Mz+Ml)/(Mz+Ml) = RzDz/RlDl \qquad 3.9$$

Thus, formula 3.9 makes it possible to calculate in the case of a mobile center of force through the values of the radii of the orbits, as if the center itself were motionless.

FORCE INTERACTION

As already noted, the concepts of the central force and the observed body are purely conventional. This is mainly due to the difference in their masses and the magnitude of the change in the dynamics of movement. In reality, it makes sense to talk only about the interaction of equal physical bodies through their gravitational or other fields. At the same time, the trajectories of movement and dynamics of bodies are a consequence of the force interaction of their fields, which always tends to balance, both from the side of the first body to the second, and from the side of the second body to the first. Their nature of interaction must necessarily have the same regularity, and its absolute value must have the same value. Then, based on 2.7 for two centers of force, we can write, 4.1:

$$(1+Dc2/Dgr)[(D'c2-D'c2)/Dc2]\ m1\ \Delta D1+m1D1 =$$
$$= (1+Dc1/Dgr)[(D'c1-D'c1)/Dc1]\ m2\ \Delta D2 + m2D2$$

or shortly, using the functional notation for dynamics:

$$K(D)\ m1(\Delta D1+D1) = K(D)\ m2(\Delta D2+D2) \qquad 4.2$$

When balancing the force interaction or when it has stopped at some level, and the bodies begin uniform rectilinear motion:

$$\Delta D1 = \Delta D1 = 0,$$

we have:

$$m1D1 = m2D2 \qquad 4.3$$

Thus, for the equilibrium state of the force interaction of rectilinear and circular uniform motions, respectively, the following conditions are met:

$$mD = Const \qquad 4.4$$

$$mDR^2 = Const \qquad 4.5$$

It is quite clear that for a rectilinear uniform motion, it is possible to speak about the equilibrium state of the force interaction only conditionally. This is the observed result of the previous force balancing, which could be either complete or partial, and the value of mD characterizes the level at which this interaction ceased. But this is a relative calculated level, since, without having data on the interaction process itself, nothing can be said about the absolute value of the centers of forces. Their meaning will be different, depending on the choice of the reference point. And only a circular motion allows to carry out the spatial binding of the two centers of forces and determine their absolute values, relative to the state of complete force balancing. The unity of these movements lies in the fact that both are the result of balancing the force interaction, and therefore are determined by the same ratios.

Another point becomes clear. Why two centers of force **M1 M2**, which move with the same dynamics **Dc1 = Dc2**, interact as if they were motionless. Let's consider two coordinate systems – **X'Y'Z'**, which moves uniformly with dynamics Dκ and motionless system **XYZ**. In each of them there are two bodies of equal mass **m1=m2** at rest, fig. 6.

Then, based on 4.1, the force interaction of these two bodies, as two centers of force, will be determined: in the fixed **XYZ** system:

$$\textbf{(D) m1}(\Delta \textbf{D1+D1}) = \textbf{K(D) m2}(\Delta \textbf{D2+D2}) \qquad 4.6$$

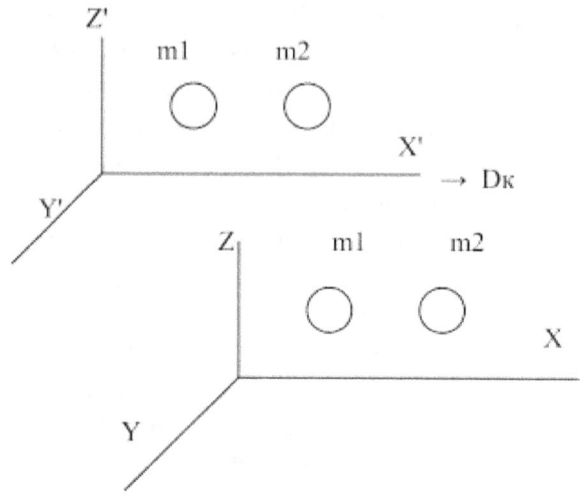

Picture 6.

For the mobile system **X'Y'Z'**, respectively, it will be:

$$\textbf{K'(D) m1}\Delta\textbf{D1+ m1(D1 + Dc) =}$$
$$\textbf{= K'(D) m2}\Delta\textbf{D2+ m2(D2 + Dc)} \qquad 4.7$$

Since, **m1Dk = m2Dk**, then equation 4.6 corresponds to 4.7 and, therefore, the observer will not fix any differences associated with the movement of systems.

If we now mentally abstract ourselves from the reference systems themselves and take the bodies in **XYZ** as motionless and consider the movement of bodies in the mobile system **X'Y'Z'** relative to them, then we can say that until the moment of observation, these bodies were in some kind of force interaction with another center of power. However, with respect to themselves, their interaction, they remained equivalent to the original state. Here it is important to emphasize the following. There is not just a formal transition from one system to another, by taking into account their dynamics of relative motion, but this is always preceded by a real force interaction.

A completely different picture will be if the masses of the bodies are not equal to each other. The formal transition from the

XYZ system to the mobile **X'Y'Z'** system would seem to violate the pattern, since **m1Dk ≠ m2Dk**. This means that the course of physical processes in various inertial reference systems (ISR) has, in fact, some differences. We will assume that the transfer of bodies from one system to another, as in the first case, was accompanied by the same force interaction up to the level mD. Then their dynamics of movement, as a result of this, will be determined as:

$$D1 = m/m1 \times D \quad D2 = m/m2 \times D$$

Thus, the transition from one system to another, in the general case, for each of the bodies has an individual character. And, if the observer first fixes in the **XYZ** system that the dynamics of the movement of bodies are equivalent: **D1=D2**, and then registers the same result in the **X'Y'Z'** system, equal to:

$$D1 + Dk = D2 + Dk,$$

then this means that between these two dimensions there was a force interaction of these bodies, with some center of force, to a different level and values of their dynamics. The difference between these levels of the speaker just determines the discrepancy between the ratios 4.6 and 4.7. in case **m1 ≠ m2.**

An analogous result is also obtained in the case when the bodies have the same value of dynamics relative to themselves, and the transition from system to system is associated with the formal consideration of the dynamics of the IFR itself. In the first case, it was about a real force interaction, however, with such a transition, there is no force interaction. Isn't this related to the very concept of an inertial system? The basis of any **ISO** is a uniformly moving reference body, which has some physical mass. And, if one reference body **ISO1** moves with dynamics **D1**, and the reference body **ISO2** moves with dynamics **D2**, then this means that, before the observer took them for reference bodies, they themselves were in some kind of force interaction, each before its own driving dynamics level: **M1 D1** and **M2 D2**. By counting from them, the observer, in essence, also counts from different levels of force interaction.

Since the process of counting the dynamics of the body's motion relative to the IFR does not imply a force interaction between the observed body and the reference body of the IFR, the mass of the reference body involuntarily escapes attention, the reference level itself is set only by the magnitude of its motion dynamics. In the future, considering the force interactions with respect to these dynamics, we are faced with the fact that their **MD** levels are different, due to the difference in masses.

Consider two bodies of mass m1 and m2 with dynamics **D1** and **D2**, then we will have:

$$\text{in ISO1:} \quad \mathbf{m1(D1 \pm Diso1); \; m2(D2 \pm Diso1)};$$

$$\text{in ISO2:} \quad \mathbf{m1(D1 \pm Diso2); \; m2(D2 \pm Diso2)};$$

It can be seen that the transition from **ISO1** to **ISO2** has a different level for bodies:

$$\mathbf{\Delta 1 = m1(Diso2 \pm Diso1)}; \qquad \mathbf{\Delta 2 = m2(Diso2 \pm Diso1)};$$

The difference in their values just indicates the force interaction. But is it fair? After all, in reality, there was no effect on the bodies. Thus, having once excluded from the analysis one of the physical characteristics, such as mass, and introduced the concept of an inertial frame of reference, we encountered a discrepancy in determining the levels of force interactions. Taking into account the absolute nature of the magnitude of the dynamics, it becomes clear that it makes no sense to talk about IFRs and their multitude. Moreover, the change in body dynamics will have a different value in different ISOs. This is clearly seen in the following example. Let two bodies of equal mass move uniformly in **XYZ** and **X'Y'Z'** systems, fig. 7. The body is simultaneously affected by the center of force M, which we will consider immovable. Then the expression for the force interaction in both systems will be:

$$\text{in the XYZ system:} \quad \mathbf{F = K(D) \; m(\Delta D + D_T)}$$

$$\text{in the X'Y'Z' system:} \quad \mathbf{F' = K(D) \; m(\Delta D' + D'm)}$$

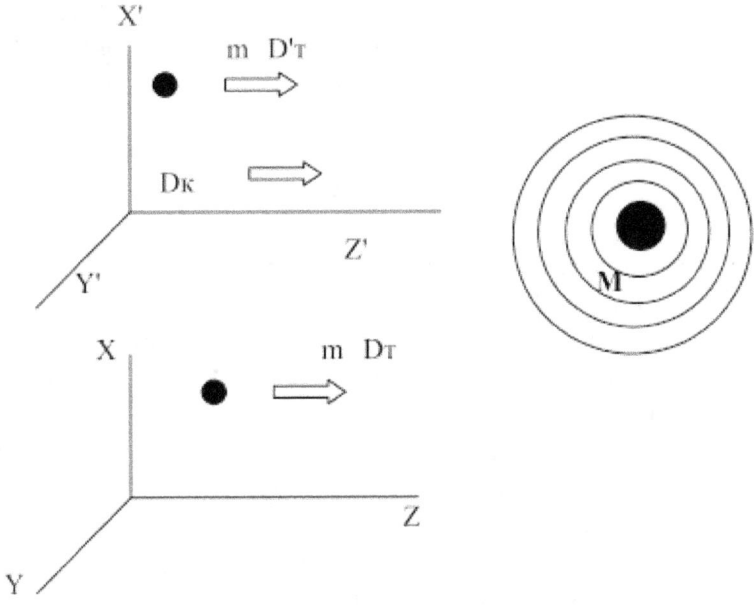

Picture 7.

where $D_T' = D_T \pm D_K$.

It can be seen that the expressions are invariant with respect to each other, therefore, the general pattern of force interaction in both systems will be preserved. However, the magnitude of the change in dynamics will be different. This is a direct conse-quence of the fact that the dynamics of body movement has its own absolute reference. If this were not so, then, since we have the same center of force M, therefore, the movements of bodies in these systems should be determined by the same characteris-tics, and hence by equal values of the forces **F** and **F '** . Using relation 2.5 we have:

$$\Delta D = (D_c - D_T')/(D_c - D_T)\, \Delta D' \pm D_K/(1\text{-}\text{ß})$$

where:

$$\text{ß} = D_t / D_c$$

The bodies, indeed, will have different values of ΔD and $\Delta D'$, which will coincide at $D\kappa = 0$. Formally, it is possible that the center of power has moved into the **X'Y'Z'** system with its own dynamics:

$$Dc' = Dc \pm D\kappa$$

which is equivalent to changing its absolute value in relation to the body. This will no doubt lead to a different value of $\Delta D'$.

It is also possible to assume that both systems are motionless relative to each other, but the body in the **X'Y'Z'** system has initial dynamics $D0 = Dk$. Naturally, in these systems, physical processes must proceed according to the same law. But what will be the difference? It is determined only by the value of **D0**. In reality, in space there is a center of force and two bodies in different systems, each of which has its own initial dynamics. And it is absolutely not necessary to indicate that one body is in a fixed inertial system, and the other is in a moving one. It is only important to know in what relation their initial dynamics are with the dynamics of force. Or, given the conventionality of their names, we can say that the result of the force interaction depends on the ratio of the absolute values of the dynamics of the movement of the centers of forces and their characteristics **M** and **D**.

ON SOME EXPERIMENTS

The most indicative, from the point of view of force interaction, is the effect of deflection of a light beam in the gravitational field of the Sun. Newton's theory of gravity, as well as Einstein's special theory of relativity, predicted roughly the same deflection result of 0.84". (10, 11). But numerous experiments that were carried out, both with light rays and in the radio range, confirmed the conclusions of the general theory of relativity (GR), which predicted twice the result of 1.75 ". (10, 11, 12, 13). At the same time, it was especially emphasized that the experiment confirms theoretical calculations with an accuracy of 2-3%.

According to modern concepts, the gravitational field is a universal force field. In turn, the electromagnetic field is one of the forms of existence of matter, which can correspond to its own measure of inertia and motion dynamics. Consequently, the expressions found earlier for the force interaction should be valid in this case as well.

Using expression 2.7 and taking into account that the light rays pass at some distance from the Sun, we get:

$$D = \frac{(Rc/R)^2 Dc^2}{D0 + (Rc/R)Dc} + D0 = D0 + \Delta D$$

where **Rc** is the radius of the Sun;

R is the distance from the center of the Sun to the light beam.

From fig. 8 shows that the deviation will be:

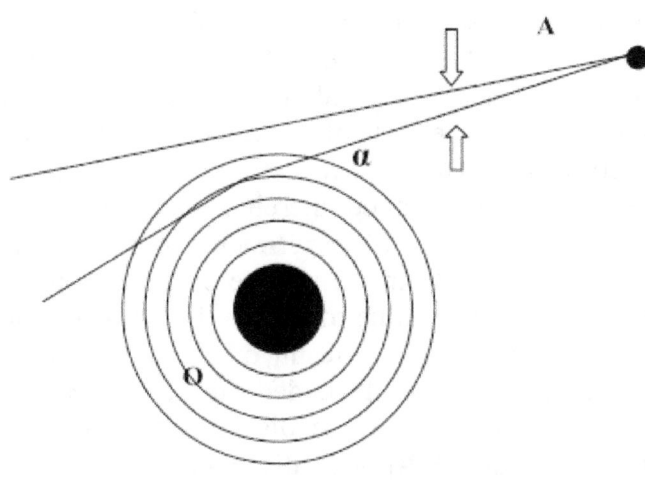

Picture 8.

$$\Psi = 2\alpha = 2\sqrt{1 - [D0/(D0 + \Delta D)]^2} \qquad 5.1$$

This takes into account that a change in the dynamics of the propagation of a light beam, and hence its deviation from a rectilinear trajectory, occurs both when moving towards the Sun and away from it. It is well known from the general theory of relativity that the final formula for calculating the deflection of a light beam in the gravitational field of the Sun has the form:

$$\Psi oto = 1.75(Rc/R) \qquad 5.2$$

Let us set **Rc = R**, while the dynamics of the Sun, as the center of force, will be:

$$Dc = \sqrt{214,9} \quad = 14,6$$

Since the calculation was made on the assumption that the dynamics of the Earth in orbit is equal to unity, then, consequently, the dynamics of light will be:

$$D0 = 3x10/29.8 = 10.$$

Substituting these values in 5.1, we find that the angle of deflection of the light beam at the surface of the Sun will be 1.486", which is 0.264" less than predicted by general relativity. As a percentage, this will be 15% of the value of 1.75". It would seem that the error significantly exceeds the error of experiments and the calculation by our method, thereby, is refuted. But it is not so. It should be noted that all experiments were carried out at a considerable distance from the surface of the Sun and, above all, in the radio range. The value of the maximum deviation at **R=Rc** was always determined as a result of extrapolation of measurement data and, of course, depended on the chosen method. The calculation shows that at these distances the numerical values according to 5.1 and 5.2 practically coincide and do not exceed the experimental error.

$$\Delta\Psi = \Psi oto - \Psi \leq 0.1"$$

On the other hand, noting the greater accuracy of the coincidence of the extrapolated value of the maximum deviation of the light beam, during the experiment, with the calculated value given by the theory, however, they are silent about significant discrepancies in the data from one experiment to another:

$$(1.57 \pm 0.08) \leq \Psi oto \leq (1.87 \pm 0.3)$$

This did not take into account the effects associated with the passage of a light beam, or rather, radio waves through the solar plasma. The theory of relativity, both SRT and GR, were very popular in their time, among scientists and the public. Obtaining the «necessary» measurement result, which would correspond to the calculated ones, was highly desirable, attracted great attention and gave fame. Therefore, the desire of scientists to present their results in a favorable light was quite understandable.

Thus, we can assume a smaller effect of deflection of the light beam 1.486" than predicted by GR. One can only add that the calculations by our method take into account the influence of only the mass of the Sun itself. But do we know this mass absolutely exactly, or is the mass of the Sun still less, and maybe more, than it is considered so far?

Another gravitational effect predicted by GR is a shift in Mercury's perigee. It has the richest history, since it is associated with the use of various approaches and views on the concept of force interaction and its mathematical description. (14). This caused a lot of controversy and discussion for a whole decade. GR explains this effect as a result of the curvature of space-time near the Sun. There were also attempts to explain this deviation due to the non-spherical shape of the Sun. (15). But at the same time, problems arose due to the difficulties of visual observation and evaluation of this non-sphericity. It was assumed that if non-sphericity exists, then it is insignificant and can explain only part of the effect. (12).

However, the velocity of the Sun in the space of the galaxy is estimated at 250 km/sec, which, according to 2.7, in itself, should significantly affect the sphericity of its gravitational field. Not the shape of the Sun itself, as a cosmic body, is non-spherical, but non-sphericity is caused, due to its movement in space, only by its gravitational field. Therefore, regardless of whether the Sun itself has a non-sphericity, its gravitational field will already be asymmetric and may lead to similar effects or make an additional contribution to the final result.

The proposed approach to the concept of force interaction makes it possible to explain the observed irregularities in the Earth's rotation from same positions. (28). Currently, this phenomenon is associated with the displacement of air masses and earth tides. (16, 29). There are even calculations based on world wind maps, which exactly correspond to the value of rotation unevenness equal to 1 microsecond, which, in itself, arouses suspicion. (30).

But all these hypotheses and calculations are good only until the moment when similar effects are registered on other planets, with other characteristics of the atmosphere or without them at all, as, say, the Moon. Such irregularities in rotation can only be an inevitable consequence of changing the conditions of force interaction between the centers of forces. When moving in space, their gravitational field is distorted: it is compressed in the direction of movement and vice versa is stretched from the opposite side. In practice, all observations are associated with a massive center of force and a small body that the observer follows. At the same time, the force center drags a small body in its circular orbit, but at the same time introduces perturbations into its movement, due to the deformation of its gravitational field.

A brief review of the observed effects and experimental data is given with the sole purpose of showing the simplicity and universality of the proposed approach to the concept of force interaction. At the same time, its simplicity and physicality play an exceptional role.

CONCLUSION

In the theory of relativity, in order to describe the mechanical motion in accordance with the experiment, it was necessary to introduce additional premises and use a cumbersome mathematical apparatus. Its correlations and conclusions led to a serious revision of the entire worldview. Disputes about the physical adequacy of the conclusions of the theory of SRT and GR do not subside even in our time.

The refusal to use the concept of time and the introduction of such a quantity as the dynamics of the relative motion of the body made it possible to use a smaller number of initial premises, to connect rectilinear and circular motion, accelerating force and momentum. We use the concept of a field extensively and distinguish them by their quality and characteristics. But what is a field, how it arose and what are its physical foundations, we still do not know. We do not know where and why the electron has an electric field or a negative charge. We do not know the nature of the gravitational field and why it is so universal.

But we have the only opportunity to study the interaction of these fields through the movement of their carriers. Therefore, the concept of the interaction of two centers of forces is universal and offers a simple and reliable way to calculate their movement. Through this concept, the contours of a single process of nature were really outlined – balancing the force interaction through the movement of matter.

LITERATURE

IN AND. Ivashov "Physical foundations of mechanical motion and its description". 1985

V.A. Ugarov "Special Theory of Relativity". M. Science, 1977

Ya.G. Dorfman. "World History of Physics". M. Science, 1974

A.K. Maneev "Continuity in the development of the categories of space, time and movement." Minsk, Science and technology, 1971

Aristotle "Physics" Gos.sots.ekon. ed. 1937

I.N. Veselovsky "Essays on the history of theoretical mechanics" M. Higher school, 1974

P.S. Kudryavtsev "History of Physics" M. State.uch.ped.giz. 1956

M. Gliozzi "History of Physics" M. Mir, 1970

G.E. Timmerding "Laws of falling bodies" Berlin, Scientific Thought. 1925

S.I. Vavilov "Experimental foundations of the theory of relativity" M. Gos.izd. 1928

S. Weinberg "Gravity and Cosmology" M. Mir, 1975

N.P. Konoplev "Gravity experiments in space" UFN 1977, v.123, issue 4

V.N. Rudenko "Relativistic experiments in a gravitational field" UFN, 1978, v.126, issue 3

N.T. Rosever "Mercury Perigee" M. Mir, 1985

R. Dicke "Gravity and the Universe" M. Mir, 1972

APPLICATION

THE LAW OF UNIVERSAL GRAVITATION.
ERROR OF I. NEWTON

I. Newton was always distinguished by a wide range of scientific and religious interests. His personal library consisted of more than 2,000 volumes in various branches of knowledge. This allowed him to unite not only the mathematical and empirical methods of antiquity, but also to raise the theory of gravitation, the cognizability of the cosmos to a new level, and change the whole idea of the world. The universality of his views and discoveries were so unexpected that it took more than one decade until they firmly formed the basis of modern science of that time.

From the vast material of his main work, The Mathematical Principles of Natural Philosophy, time has preserved and crystallized for us only his basic laws of dynamics. His geometric methods and constructions are more reminiscent of a collection of solutions to original problems in geometry, moreover, he formulated the initial conditions of these problems himself. Those who dare to read and deal with every decision, statement, lemma or corollary will very soon become disheartened. Even in Newton's time, Cambridge students jokingly said, "Here comes a man who has written a book in which neither he nor anyone else understands anything."

But why exactly did the geometrical method attract Newton's attention? He himself answers:

«So, geometry is based on mechanical practice and is nothing but that part of general mechanics in which the art of precise measurements is expounded and proved».

But in this vast pattern of geometric constructions, there are the greatest thoughts of the scientist and his views on the basic

145

concepts of physics such as time, space, gravity. And his laws of dynamics made it possible to break out of the Earth into space and calculate the main parameters of the planets of the solar system. Newton argued that gravity is a general property of bodies and extends far beyond their limits. This is the real reason, not some hidden property.

Interestingly, Newton distinguishes between general mechanics and rational mechanics:

«In this sense, rational mechanics is the doctrine of the movements produced by any forces, and of the forces required for the production of any movements, precisely stated and proved».

It was a grandiose breakthrough into the future, which brought a new worldview of all mankind.

Many ideas of physics were expressed even before Newton, but it was he who managed to lay the foundation for a new classical physics. Therefore, let us consider its main ideas and laws, which combined the experiment and the theory of the physics of the motion of bodies. Some argue that Newton did not give a mathematical formulation of his laws in his book. Indeed, he wrote down the laws themselves in writing, but mathematically adequately. This did not prevent him from using these laws when calculating the motion of bodies and planets. In the preface to the book, Newton wrote: "This essay means a careful development of the application of mathematics to physics." The book begins with definitions as the main foundation for further research. In them, Newton tried to fill with physical meaning those foundations that were previously perceived as some abstractions.

Definition No. 1.

«The quantity of matter (mass) is a measure of that, established in proportion to its density and volume».

This quantity I mean hereinafter under the name body or mass. Mass is determined by body weight, because it is proportional to weight.

Definition No. 2.

«The amount of movement is a measure of such, set in proportion to the speed and mass».

Definition No. 3.

«The innate force of matter is its inherent innate resistance, according to which any single body, since it is left to itself, maintains its state of rest or uniform rectilinear motion».

It happens from the inertia of matter that every body is only with difficulty removed from its rest or movement. Therefore, the «innate force» could very intelligibly be called the «force of inertia».

Definition No. 4.

«An applied force is an action performed on a body in order to change its state of rest or uniform rectilinear motion».

Force manifests itself only in action and does not remain in the body after the action ceases.

Definition No. 5.

«The centripetal force is that with which bodies are attracted to a certain point, as to a center, from everywhere, driven, or somehow striving».

In the centripetal force, three kinds of quantities are distinguished: absolute, accelerating and driving.

Definition No. 6.

«The absolute value of the centripetal force is a measure of the greater or lesser power of the very source of its propagation from the center to the space surrounding it».

Definition No. 7.

«The accelerating value of the centripetal force is a measure proportional to the speed that it produces during a given time».

So, the action of the same magnet is stronger at close range, weaker at a far distance.

Definition No. 8.

«The driving magnitude of the centripetal force is a measure proportional to the amount of movement that it produces during a given time».

This value is the aspiration of the whole body directed towards the center, which is called its weight.

Very important are his philosophical teachings on the nature of the basic concepts of physics:

– Absolute, true, mathematical time in itself and in its very essence, without any relation to anything external, flows evenly and is otherwise called duration.

– Relative, apparent or ordinary time is, or exact or changeable, comprehended by the senses, external, performed through any movement, a measure of duration used in everyday life instead of true mathematical time, such as: hour, day, month year.

– Absolute space in its very essence, regardless of anything external, remains always the same and immovable.

– A place is a part of the space occupied by a body, and in relation to space it can be either absolute or relative.

– Absolute movement is the movement of a body from one absolute place to another, relative movement is from relative to relative.

– Absolute and relative movement and absolute and relative rest differ from each other: properties, causes of origin and manifestations. The property of rest is that bodies that are truly at rest are at rest relative to each other.

– The property of movement is that the parts that retain their position in relation to the whole participate in the movement of this whole.

It is only after these preliminary remarks that Newton proceeds to formulate his basic laws of motion.

Law 1. Any body continues to be held in its state of rest or uniform and rectilinear motion, until and insofar as it is forced by applied forces to change this state.

Law 2. The change in the momentum is proportional to the applied driving force and occurs in the direction of the straight line along which this force acts.

If any force produces a certain amount of motion, then a double force will produce a double, a triple force will produce a triple.

Law 3. An action always has an equal and opposite counteraction, otherwise the interaction of two bodies against each other is equal and directed in opposite directions.

I. Newton always carefully selected words and expressions for his proofs. Sometimes he rewrote one page 5-6 times. Therefore, it is necessary to read its basic laws with special attention and diligence. What ideas did he propose for considering the problems of attraction of spherical bodies?

Theorem XXXV. If equal centripetal forces are directed to individual points of a given ball, decreasing in proportion to the squares of the distances to these points, then I say that any such ball is attracted first with a force inversely proportional to the square of the distance between the centers of the balls.

In order to comprehensively clarify and deepen his thought, he deduces several consequences from this main theorem, two of which are of particular interest.

Corollary 1. The attraction of other homogeneous balls by the balls is proportional to the volumes (masses) of the attracting balls, divided by the squares of the distances of their centers to the centers of the attracted balls.

Corollary 2. The same thing happens in the case when the attracted ball itself attracts. «...in every attraction, both the attracted point and the attracting point are equally stimulated, then two mutual attractive forces will be formed that maintain the same proportion».

And, further, in his Terem XXXVI, I. Newton gives two consequences, which we already know today as the law of universal gravitation, but again without a mathematical formula.

Corollary 1. Therefore, if several balls of this kind are attracted mutually, then the accelerating forces of attraction by each separate ball of the other will be at equal distances from the center, proportional to the masses of the attracting balls.

Corollary 1a. At different distances, these forces are proportional to the masses divided by the squares of the distances to the centers.

Corollary 2. The driving forces of attraction, otherwise the weight of one ball on another, with equal distances between the centers, will be proportional to the products of the masses of the attracting and attracted ball.

Corollary 2a. At unequal distances, these forces are directly proportional to the said product of the masses and inversely proportional to the squares of the distances.

It can be seen that the corollaries of Theorems XXXV and XXXVI contradict each other. On the one hand: *«two mutual attractive forces will be formed, maintaining the same proportion»*, i.e. they are proportional to the masses of these bodies and divided by the squares of the distances. And, on the other hand, Newton claims *«that forces are directly proportional to the said product of masses and inversely proportional to the squares of distances»*. Thus, he equated the two mutual forces of attraction, which, in principle, should have been added, to the product of these forces. The statement of «Corollary 2a» later served as the main argument for commentators in compiling the general formula for universal gravitation.

Extending his principle to the attraction of the planets, he writes:

«Gravity exists for all bodies in general and is proportional to the mass of each of them ...». *«All the planets gravitate towards each other, and also that the gravitation towards each of them individually is inversely proportional to the squares of the distances of the place to the center of this planet».*

This principle formed the basis of all his calculations of the motion of the planets, their masses and orbits. In his teachings, I. Newton wrote:

«We must not accept other causes in nature, beyond those that are true and sufficient to explain phenomena».

Newton's views not only opposed the teachings of Descartes, but also went against the teachings of the church in some respects. For example, a particularly sharp dispute took place on the question of the motion of the Earth. The Church denied the very possibility of the Earth moving around the Sun and made great efforts to defile Newton's hypothesis.

Very definitely I. Newton spoke about the magnetic force.

«Gravity is of a different kind than magnetic force, because magnetic attraction is not proportional to the attracted mass: some bodies are attracted more strongly, others are weaker, most are not attracted at all. The magnetic force in the same one body can be increased and decreased, sometimes it is even much greater, referring to mass, than the force of gravity; when moving away from the magnet, it does not decrease in inverse proportion to the squares of the distances, but closer to the cubes, since, I could judge from some rough experiments».

This passage shows that at that time the doctrine of electricity and magnetism was in its most infancy. The basis of physics was the mechanics and kinematics of the motion of bodies.

The basic views and statements of I. Newton himself allow a deeper understanding of the level of development of physical ideas, which we gratefully use today. Even the creation of the theory of relativity did not shake this foundation. It should be noted that I. Newton in his research always proceeded and relied on the data of the experiment. And only then he generalized this experience and deduced laws. So in the case of the concept of accelerating force, he used the data of Galileo's experiments with the fall of heavy bodies. Galileo determined the main principle that all bodies, regardless of weight, equally reach the surface of the earth, i.e. have the same acceleration, under the influence of the force of gravity of the earth. I. Newton generalized this principle to all bodies and distances. He reasoned like this:

«Let's denote the rate of change of some value X, the same letter, but with a dot at the top **X'**. Momentum through **P**, and the rate of change of momentum through **P'**. Force through F, then: **P'=F, P=mV**».

Since Newton's mass does not change, therefore, the speed changes: **P'=mV'**. But **V'=a** is called acceleration.

From here, his famous formula was presented by commentators:

$$F = ma.$$

Thus, it should be specially noted that Newton is not talking about a static force, as in Galileo, but about a change in force during an accelerated motion of a body. And this change of force acts along the line of motion of the body.

He extends Galileo's private law to the whole variety of forces, masses and accelerations. Indeed, Galileo could use only one attractive force of the Earth. In relation to the size of the Earth and its weight, his trial balloons were nothing more than point bodies. So he talked about constant force and constant acceleration for his particular experience. But Newton asserted the relation not about a constant force, but about a change in the momentum of the force, throughout the movement of a body of a certain mass. Then his law would have to be written:

$$\Delta F=m\Delta V/\Delta t \text{ or } \Delta F=ma.$$

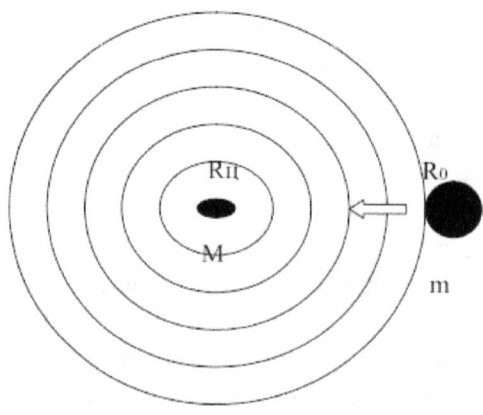

Picture 1.

If we move on to the concept of the speed of relative movement – dynamics, then this formula will take the form:

$$\Delta F = m\Delta D$$

With each transition of the body from one radius to another, the bending force and, accordingly, the dynamics of the body increases.

At each radius, the magnitude of the force will be proportional to $FR = M/R^2$, which can be denoted as the gravitational potential Ψn.

Then ΔF will be the difference between these two gravitational potentials:

$$\Psi n - \Psi n\text{-}1 = m\Delta D$$

At the same time, it is necessary to understand well that from a physical point of view, Rц can never be equal to zero. It can only have a minimum value equal to the radius of the body itself. The equality $R = 0$ means the absence, the disappearance of the body of the central force. In this case, there can be no force interaction. Then, in the general case, it is necessary to write:

$$F = M/(Rn + Rc)^2 = m\Delta D$$

where Rc is the radius of the body itself, and Rn is any radius in the space of the force field.

Hooke, at the beginning of 1680, writes to I. Newton that he assumes that the force of attraction is inversely proportional to the square of the distance between the centers of the planets and the Sun. Newton himself wrote earlier:

«In the same year (1666) I began to think about gravity extending to the orbit of the Moon, and found how to estimate the force with which a ball rotating around a sphere presses on the surface of this sphere. From Kepler's rules that the periods of the planets are in one and a half proportion to the distance of the centers of their orbits, I deduced that the forces holding the planets in their orbits must be in the inverse ratio of the squares of their distances from the centers, around which they revolve.

From here I compared the force required to keep the Moon in its orbit with the force of gravity on the surface of the Earth and found that they almost corresponded to each other».

Note that Newton, at the beginning of his reflections, speaks of the force that acts on the planets and keeps them in their orbits. Based on this, he comes to the original formula for the force action of the center of force:

$$F = \mu M/R^2$$

where μ is the coefficient of proportionality.

Investigating, further, the laws of motion of bodies, he comes to the final conclusion:

«Gravity exists towards all bodies in general and is proportional to the mass of each of them ... and inversely proportional to the squares of the distances to their places».

Let us pay special attention to the fact that I. Newton speaks of proportionality to the masses of bodies, but does not reveal anything from how these masses are interconnected. He does not provide a definitive solution to this problem, nor does he give a mathematical expression for the law of gravity, which is widely known to us today. Commentators did their best for him. Following their logic, they assigned the proportionality coefficient μ the value of the mass of the second body. Thus, I. Newton's law of gravity acquired its final form:

$$F = Mm/R^2$$

Subsequently, only one more coefficient of proportionality was added, which was called the constant of gravity.

The force interaction of two bodies became proportional to the product of their masses. But on what basis? Let's look again at Fig.1. It conventionally depicts only the gravitational field of the central force. But the body itself has a gravitational field. Suppose that both bodies approach each other until their outer shells touch. In this case, the maximum force of their mutual at-

154

traction will be observed, proportional to the product of their masses. What will happen in their environment?

Mentally combine both bodies into one whole, with a total mass of **M + m**. There will arise a central force proportional to this sum of the masses of both bodies, but not their products. Since the mutual attraction of two bodies occurs through their gravitational fields, then these fields are actually multiplied, and not the masses of the bodies. Is it possible to multiply the values of two fields? Apparently not.

The mutual attraction of two bodies occurs through their gravitational fields, which penetrate each other and add up. It has already been proved more than once that fields cannot multiply, but only add or overlap. We say that these fields are proportional to the masses, but then the magnitude of their interaction is inevitably proportional to their sum. At the time of Newton, of course, the doctrine of fields was not known, and therefore such a distortion of the physical principle appeared.

How should I. Newton's law look from the point of view of our modern views on the nature of the interaction of fields:

$$F = \gamma(M+m)/R^2 \qquad 1.2$$

where γ is the gravitational constant.

Note that, in this case, the magnitude of the force interaction corresponds to and is proportional to the values of the two gravitational potentials of both bodies:

$$M/R^2 \text{ and } m/R^2$$

This circumstance unites the whole doctrine of the force interaction of bodies. Expression 1.2 is valid only for the equilibrium state of two bodies, i.e. when the bodies are either stationary or rotate around their common center.

For the law to acquire a dynamic character, i.e. bodies acquired mutual acceleration, with necessity there should be a change in the interacting force in the direction of its increase. Since the masses of the bodies do not change, then the distance between the bodies changes, i.e. decrease in the distance between the centers of the bodies:

$$\Delta F = \gamma(M+m)/\Delta R^2 \qquad 1.3$$

In this case, ΔR cannot be less than the sum of the radii of both bodies.

This law also perfectly describes the motion of the planets in their orbits, but only their masses require a certain correction. However, how were these masses calculated? All calculations proceeded from an approximate determination by the methods of geodetic measurements of the value of the mass of the Earth. Do we know this mass for certain? Obviously, no. Further, the distances to the Sun were determined and its mass was calculated. This was the basis, which was based on I. Newton's law of gravity. This is where it becomes clear why I. Newton wrote in his remarks that the forces are directly proportional to the said product of the masses and inversely proportional to the squares of the distances. The calculation of the mass of the Sun looked like the equality of force and acceleration. The force of attraction between the Earth and the Sun is equal to the centripetal force.

Centripetal force: $F = mv^2 / r$. Force of attraction: $F = GmM / r^2$. Hence, comparing the two values, we get:

$$mv^2 / r = GmM / r^2$$

It is clearly seen that the mass of the Earth «disappears» in the calculation. In reality, it turns out that its gravitational field does not take any interaction with the field of the Sun. The earth has turned into a «trial ball» and this ball is moving in the field of the central force. Hence the mass of the Sun, as a central force, is calculated as a purely mathematical example, but only for the Earth's orbit:

$$M = r v^2 / G = 1.5 \times 10^{11} \times (3 \times 10^{\wedge}4)^2 / 6.6725 \times 10^{11} =$$
$$= 2 \times 10^{30} \text{ kg}$$

There was another paradox as well. If we claim that this is the law of universal gravitation, then all such calculations should be valid for other planets, as well as their satellites, i.e. in general for all bodies. This is the law! Therefore, knowing the mass of the Sun, it is possible to calculate the mass of any planet. However, surprisingly, the masses of the rest of the planets were found in other ways. The determination of the masses of planets with satellites was carried out on the basis of Kepler's III law. The masses of Mercury and Venus were originally determined by the perturbations they cause in the motion of other planets. Flights to these planets of spacecraft made it possible to clarify the values of their masses, according to their effect on the trajectory of the apparatus. The mass of Pluto could, approximately, be determined only after its satellite was discovered. All values of the masses of other planets, except for the Sun itself, were found using astronomical methods and measurements. Why so? It's just that the law of universal gravitation does not give the same value of the mass of the Sun when calculating, through the masses of other planets. It turns out that each planet has its own «personal» Sun. Let's see what mass the Sun has, for example, for Saturn:

$$M = r\,v^2 / G = 1.427 \times 10^{\wedge}12 \times (9.69 \times 10^{\wedge}3)^2 / 6.6725 \times 10^{11}$$
$$= 2.073 \times 10^{\wedge}29 \text{ kg}$$

Same for Mars:

$$M = r\,v^2 / G = 227.9 \times 10^{\wedge}9 \times (24.13 \times 10^{\wedge}3)^2 / 6.6725 \times 10^{11}$$
$$= 3.9 \times 10^{\wedge}29 \text{ kg}$$

At the same time, one should not forget that all masses were calculated for point-like «balls-planets» that abstractly rotate in their orbits.

Imagine mentally that we are observing our solar system from the side and can see the gravitational fields of the planets and the Sun in the form of a bluish glow. A powerful glow from the Sun spreads throughout space. It is not uniform and decreases, moving away from it, according to a well-known law – inversely proportional to the square of the distance. Inside this glow, planets float in their orbits, which complement the solar glow with their own, like luminous balls. The distribution of this glow also obeys the law inversely proportional to the square of the distance from their centers. All these glows of the planets complement and add up with the solar glow, as well as with each other. An intriguing picture of a light carousel. But how did all these planets end up inside the glow of a huge solar ball or inside its gravitational field? Let's assume that they flew past and were captured by the gravitational field of the Sun. Each planet took its place depending on its mass and speed in the solar orbit. Thus, the external gravitational field of the Sun no longer affects the acceleration of the planets, and therefore their attraction. The determining factor is the internal, which is located under the sphere, limited by a radius equal to the trajectory of the planet. But in this case, the gravitational potential of this supra will be determined only by the volume of the gravitational field that is under this sphere. Then the formula for the gravitational potential will be written as:

$$\Psi = (3M/4\,\pi R^3)/4\,\pi R^2 = 3M/R$$

where R is the distance between the spheres of the Sun and the planet. It has already been said that the mutual attraction and acceleration of the planets in the field of the Sun can only occur until they come into contact at the level of the spheres, then their collapse occurs.

Is there a gravitational field inside the body itself? Of course there is, it is its source.

However, a decrease in the radius inside the body will lead to a loss of mass and a decrease in the gravitational field itself. The greatest gravitational field is observed only on the very surface of the body. Therefore, taking into account the radii of the planets and the Sun, the law of universal gravitation, in general, will take the form:

$$F = \gamma[3M/(Rc + Rn) + 3m/(Rp + Rn)] \qquad 1.4$$

M is the mass of the Sun; m is the mass of the planet; Rc is the radius of the Sun; Rp is the radius of the planet; Rn is the distance between the spheres of the Sun and the planet. Since the radii of the planets are many times smaller than the distances to the Sun, we can simply write:

$$F = \gamma[3M/Rn + 3m/Rn] = 3\gamma[(M +m)/ Rn] \qquad 1.5$$

In the case of calculating centripetal acceleration, it is also necessary to take the distance between the spheres of the Sun and the planets. We repeat that gravitational fields interact, not the masses of bodies. Then, to determine the mass of the Sun, we will have:

$$3\gamma[(M+m)/ Rn] = mv^2 / Rn$$

$$Mc = mv^2/3\gamma \qquad 1.6$$

To calculate the mass of the Sun, it is necessary to use, as the most reliable measurement data, the mass and orbital velocity of the Earth. And already through the known mass of the Sun to determine the masses of the planets of the entire solar system. The data of these calculations are given in Table 1.

The first and second columns present the speed and mass data of the planets, which are accepted today. In the third column, the mass of the Sun, calculated according to 1.6, and the masses of the planets, calculated using the same formula, but with a known mass of the Sun.

It is clearly seen from the table that the Sun is a much more massive star than has been accepted so far. For the planets, Jupiter and Pluto have the greatest discrepancies. It should be especially noted that all calculations were made taking into account the magnitude of the masses of the Sun and planets, as well as the speed of the planets in orbit. Thus, there are all the components that really affect the physical process of the gravitational attraction of the Sun and planets.

TABLE 1

	km / sec	kg	kg
SUN		2x10^30	2,67x10^43
Mercury	47,87	3,3x10^23	2,32x10^24
Venus	35,02	4,9x10^24	4,36x10^24
Earth	29,78	6x10^24	6,03x10^24
Mars	24,13	6,4x10^23	9,18x10^24
Jupiter	13,07	1,9x10^27	3,14x10^25
Saturn	9,69	5,68x10^26	5,68x10^25
Uranium	6,81	8,69x10^25	1,15x10^26
Neptune	5,43	1,02x10^26	1,81x10^26
Pluto	4,67	1,303x10^22	2,45x10^26

The most surprising result is the calculation of the mass of the moon. Today it is accepted that the mass of the Moon is 7.35x10^22 kg, its speed is 10^3 m/sec. Calculation according to formula 1.6 gives a result of only 1.2x10 ^ 9 kg. Perhaps the answer lies in the fact that the Moon is an artificial body and there are large cavities inside it. In this case, it should be borne in mind that the Moon always faces the Earth on one side and does not rotate with respect to the Earth. She seems to be tied with a very strong thread.

What could be causing this? Only at the expense the fact that the mass and density of the Moon is much greater on the side that faces the Earth. On the reverse side there are large voids. Giving free rein to fantasy, we can say that all kinds of technical devices are located on the side of the Earth: energy sources, special emitters and receivers for research and observation of the inhabitants of the Earth, and possibly also engines if it is a spacecraft. On the reverse side – premises and laboratories of aliens, as well as a fleet of their small spacecraft. In this case, we know very little about the Moon and interesting discoveries await us.

Not only the Moon, but the Earth itself, is a planet full of mysteries, which also suggests external artificial influence during its creation. Literally all aspects of its structure cause numerous questions. For example: the presence of water, the presence of an atmosphere, a protective magnetic field, the optimal location of the orbit for human life, amazing flora and fauna, as well as the very appearance of man. One can only smile at the hypothesis of the origin of man from apes. Mankind has received unique sources of knowledge about how man was created and from where he was brought to Earth. First of all, these are the texts of the Egyptian pyramids, which are set forth in the "Book of the Dead", and, secondly, these are the Five Books of the Bible (Torah). The secret story tells about the artificial creation of a person in the process of a genetic experiment on a distant planet of androgynes. It is the androgynes, as superconscious beings, who are our Creators and creators. Ancient people called them Gods and Demigods. After the revival of humanity of the second generation, the Gods flew back, but the Demigods remained and created their androgynous civilization. They exist in parallel with us! (Ivi Solpe «The Androgynous Creation That the Bible Hid» 2013, Ivi Solpe «The Living Word of the Book of the Dead» 2015).

ANNOUNCEMENT

The book contains three articles by the author dedicated to the fundamental concepts of modern physics and our worldview: time, force interaction, coordinate system.

Much attention is paid to the historical aspect of how people's ideas arose in the process of their labor activity. First of all, it is concluded that the concept of time was introduced into the description of mechanical movement artifi-cially, based on the practical experience of people, the need to streamline their daily life and work. Gradually, the concept of time became a fundamental physical quantity. It is included in all theoretical descriptions of physical processes, such as speed, acceleration, and manifests itself in the coordinate system, as a seg-ment of the path, etc. But is time a real physical substance or cotinuum? If not, how should our idea of the world and the possibility of describing it look like? The author gives an unambiguous answer – time, as a physical reality, does not exist! Based on the principle of relativity, a new concept of «body dynamics» is introduced as a number.

Descartes's coordinate system is also the heir to the practical experience of people who perceived the earth's surface as flat. They did not see themselves as the «observer» of the world from the center of the system. For them, there always existed some particular movement of a material object in one direction, which they tried to describe. However, the Michelson-Morley experiment made its own adjustments. Time has become variable, depending on speed. As a result, the distance has become changeable. The formulas for the transition from one coordinate system to another have also become much more complicated. What did the Michelson-Morley experiment prove? Was Einshtein right when he introduced the postulate of the constancy of the speed of light and created the special theory of relativity? The experiment showed that our ideas about the coordinate system are wrong.

The author offers a completely new view of the coordinate system as radial-spherical. In this system, the Michelson-Morley experiment does not have a «negative» result and is described as the usual propagation of light in all directions, with the same speed.

Having discovered the law of attraction of two bodies to the world, Newton made a logical mistake, interpreting his final result of calculations. For centuries, using the formula in practical calculations and getting results that are quite suitable for practice, people do not even ask the question:

«Was Newton right?»

But a small logical analysis is enough to understand that, the force interaction of two bodies is proportional not to the product of their masses, but to their SUM. This form of the law also perfectly describes the motion of planets in the circle of the Sun, but there are also some quantitative differences. Modern methods of experimental research make it possible to identify these differences and finally answer all the main questions.

A person must finally understand his place in this world and learn how to correctly display it in theoretical descriptions.